SpringerBriefs in Applied Sciences and Technology

Forensic and Medical Bioinforma

For further volumes:
http://www.springer.com/series/11910

Amit Kumar

Biotechnology and Bioforensics

New Trends

Springer

Amit Kumar
BioAxis DNA Research Centre (P) Limited
Hyderabad
India

ISSN 2196-8845 ISSN 2196-8853 (electronic)
ISBN 978-981-287-049-0 ISBN 978-981-287-050-6 (eBook)
DOI 10.1007/978-981-287-050-6
Springer Singapore Heidelberg New York Dordrecht London

Library of Congress Control Number: 2014942057

Printed on acid-free paper

Springer is part of Springer Science+Business Media (www.springer.com)

Foreword

DNA technology has engraved its name in improving the nutritional quality of food by producing GM foods, personalized medicine based on DNA screening techniques, forensic DNA testing, genetic testing, and many other domains. New inventions and outbreaks in the era of science can be best exemplified with the progress in the implementation of DNA-based technologies and methodologies. This series presents the information on genetic predisposition, conventional communized medicine, and the individual specific problems with a focus on the therapy, often a random one rather than being specific as in the case of personalized medicine based on DNA screening techniques. The literature in this series describes the problems posed by the loss of victim identity in mass disaster cases that can be better solved by the DNA profiling technology today. It is also a well-accepted fact that Bioinformatics now plays a key role in modern drug discovery and it cuts enormous time frame and funding required in the processes of drugs reaching the market. Modern tools in bioinformatics and structure-based drug discovery have now enabled Insilco studies of crucial steps like weeding out the bad leads early in the process, positive lead screening, stability, toxicity prediction, bioavailability studies, etc. The chapters in this series comprise research insights from bioinformaticians, pharma professionals, and computational biology and discusses the emerging challenges, problems, and achievements in this field of developing algorithms and softwares, databases, target identification, Protein modeling, Protein function prediction, Binding site analysis, Protein drug complexes, Biomarkers Research Docking, etc.

Preface

The current era of incredible innovations toward the zeal to chase the heights of development has made science and technology one of the most powerful tools to accomplish the tasks of incremental prosperity for human welfare and sustainable development. It has been rightly said that science, technology, and innovation work together for growth and development. With the multifarious aspects of science there is a need for thought-provoking ideas and cumulative efforts which can strengthen the scientific capacity to produce successful innovation systems. This series has come up with its horizons widened. It aims to bring out all the advancements and innovations in the field of Biology, not restricting itself only to DNA technologies. The chapters in the book have brought together interdisciplinary domains of Life Science to implement the developments in one discipline so as to foster the standards of the other. I hope this series would be both knowledgeable and memorable to you all. We can expect more automation; it is already happening. There will be more integration of computerized analysis with laboratory tests. Capillary electrophoresis will require less material and produce faster results, DNA chips are in the pipeline too. We can also expect miniaturization with attendant portability. I recently read of a hand-held chip that can analyze 8 STRs in a few minutes. We can foresee the time when analysis can take place at the crime scene. If immediate results are produced, this can provide prompt clearance of erroneously identified suspects, avoiding much needless apprehension. I would emphasize, however, that what can be done in pilot experiments will usually not be good enough for forensic use, for which a system must be thoroughly tested and validated. While the appropriate use of DNA can be helpful in reducing and reversing wrongful convictions, inappropriate use of it and the sway of it, over other evidence on juries and judges can create a system of wrongful convictions. I offer my sincere gratitude toward the authors of the book chapters, acknowledging the support and cooperation received from the scientists from BioAxis DNA Research Centre Private Limited, CRRao AIMSCS, Andhra Pradesh Forensic Science Laboratories, Hyderabad from where most of the reviewers contributed toward review of papers.

Contents

Chapter 1
Amplification and Sequence Analysis of TPI Gene, a Structural Gene of Operon from *Lactobacillus delbrueckii*

T. Pravin Reddy and Dhatrika Sahithi

Abstract The main objective of this work is amplification and sequence analysis of TPI gene which is a structural gene of operon from *Lactobacillus delbrueckii*. The cow milk was collected under good conditions and analyzed for bacterial load. Many strains of lactic acid bacteria were obtained, out of which 3 strains were homofermentative lactobacilli. Different biochemical tests were done for the identification of *Lactobacillus delbrueckii*. The TPI gene of *Lactobacillus delbrueckii* is involved in the maximum production of bacteriocin. Genomic DNA was isolated from *Lactobacillus delbrueckii* and its TPI gene was successfully amplified using polymerase chain reaction. Further, sequence analysis of TPI gene was done using the bioinformatics tools and the phylogenetic study was made. From the sequence analysis we can infer that the TPI gene sequence of *Lactobacillus delbrueckii* has close resemblance with the *Escherichia coli* sequence.

Keywords *Lactobacillus delbrueckii* · TPI gene · Bioinformatic tools · Phylogenetic analysis

1.1 Introduction

Lactobacillus delbrueckii is a Gram-positive bacteria and a facultative anaerobe. It is long, filamentous and non-motile with its cell size ranging between 0.5–0.8 by 2.0–9.0 mm. The environment in which *L. delbrueckii* thrives the most would be one of relatively low pH, preferring acidic environments making them acid tolerant. *L. delbrueckii* undergoes obligate homofermentative metabolism which means it is only able to ferment lactose and no other sugar. *L. delbrueckii* is found in dairy products such as yogurt, milk, and cheese.

T. P. Reddy (✉) · D. Sahithi
BioAxis DNA Research Centre (P) Limited, Hyderabad 500 068,
Andhra Pradesh, India
e-mail: praveenreddy546@gmail.com

A. Kumar, *Biotechnology and Bioforensics*, Forensic and Medical Bioinformatics,
DOI: 10.1007/978-981-287-050-6_1,

In 1905, Stamen Grigorov, a Bulgarian doctor, discovered *L. delbrueckii* subspecies bulgaricus to be a true cause for the existence of natural yogurt. This was an important discovery since it led to further discovery of probiotic effects on humans and animals which improves lactose tolerance and has the ability to stimulate immune responses.

That strain of bacillus was *L. delbrueckii* subsp. Bulgaricus. A recent study reported the isolation and characterization of *L. delbrueckii* subsp. bulgaricus along with its symbiont Streptococcus thermophilus from plants in Bulgaria were the basis of traditional yogurt preparation [1]. Economic losses would be significant if the fermentation process of the widely used *Lactobacillus delbrueckii* subsp. *bulgaricus* and subsp. *lactis* were hindered. Thus, the dairy industry must be able to detect bacteriophages and adjust conditions of production to ensure high quality for safety and shelf life [2].

The circular genome of *Lactobacillus delbrueckii* subsp. *bugaricus* ATCC 11842 Composed of 1,864,998 nucleotides, it has an unusually high G-C content (49 %) in comparison to other species of the genus *Lactobacilli* to which it belongs. Of the 2,217 genes present 1,562 genes that code for proteins part B. Within the *lac* operon is the *lac*S, *lac*Z, and *lac*R genes that encodes for the uptake and breakdown of lactose [3, 4]. The *lac*S gene codes for lactose permease responsible for the ability to transport lactose in through the membrane. The important enzyme B-galactosidase necessary for lactose metabolism is encoded in the *lac*Z gene. Downstream of *lac*Z is the regulatory gene *lac*R.

TPI gene is a structural gene of operon from *Lactobacillus delbrueckii*. TPI gene is Triosephosphate isomerase gene. The physiological function of TPI is to adjust the rapid equilibrium between dihydroxyacetone phosphate and glyceraldehyde-3-phosphate produced by aldolase in glycolysis, which is interconnected to the pentose phosphate pathway and to lipid metabolism via triosephosphates.

Bacteriocins are proteinaceous toxins produced by bacteria to inhibit the growth of similar or closely related bacterial strain(s). They are typically considered to be narrow spectrum antibiotics, though this has been debated [5]. They are phenomenologically analogous to yeast and paramecium killing factors, and are structurally, functionally, and ecologically diverse. Bacteriocins were first discovered by A. Gratia in 1925 [6, 7]. One method of classification fits the bacteriocins into Class I, Class IIa/b/c, and Class III. One important and well studied class of bacteriocins is the Class IIa they are all cationic, display anti-*Listeria* activity, and kill target cells by permeabilizing the cell membrane [8–10].

Bioinformatics plays a key role in handling, integrating and analyzing the flood of ‘omics’ data being generated. Reconstruction of metabolic potential using bioinformatics tools and databases, followed by targeted experimental verification and exploration of the metabolic and regulatory network properties, are the present challenges that should lead to improved exploitation of these versatile food bacteria [11]. On the phylogenetic tree the distance between wch9901 evolution branch and *Lactobacillus delbrueckii* subsp. bulgaricus LGM2 evolution branch was the closest [12].

1.2 Materials and Method

1.2.1 Collection of Pure Cow Milk

Cow milk is collected in sterile bottles and then transported quickly to the laboratory to be analyzed. The analysis was done at bioaxis DNA Research Centre Laboratory, A Centre for Biological Research, L.B. Nagar.

1.2.2 Isolation of *Lactobacillus delbrueckii* by Direct Inoculation and Serial Dilution

Lactobacillus delbrueckii can be easily isolated by serial dilution method. 1 ml of Cow milk is drawn aseptically into test tubes in preparation for serial dilution to provide 10^{-1} which was used for further dilutions to 10^{-9}. About 4 g De Mann Rogosa and Sharpe (MRS) agar was dispensed into 75 ml distilled water and autoclaved at 121 °C for 15 min and then poured into respective Petri plates. 0.1 ml of solution from each test tube are dispensed aseptically into sterile Petri dishes by spread plate technique and incubated at 37 °C for 24 h. The most common method to isolate individual cells and produce a pure culture is to prepare a streak plate. These plates are incubated to allow the growth of colonies.

1.2.3 Gram Staining

Gram staining is an empirical method of differentiating bacterial species into two large groups (Gram-positive and Gram-negative) based on the chemical and physical properties of their cell walls. The Gram stain is almost always the first step in the identification of a bacterial organism.

1.2.4 Biochemical Confirmation

1.2.4.1 Catalase Test

Use a loop to remove little bit of bacteria from the plate and smear it on a microscope slide. Add one drop of hydrogen peroxide. If the organism produces a catalase, rapid bubbling will occur.

1.2.4.2 Glucose Fermentation Test

Take two test tubes. Add 5 ml of glucose broth to the test tubes. Place inverted Durham tubes into the test tubes. Autoclave the tubes at 121 °C for 15 min. Inoculate the test organism into one of the test tube and name it as test. Incubate overnight.

1.2.4.3 Mannitol Fermentation Test

Take two test tubes. Add 5 ml of mannitol broth to the test tubes. Place inverted Durham tubes into the test tubes. Autoclave the tubes at 121 °C for 15 min. Inoculate the test organism into one of the test tube and name it as test. Incubate overnight.

1.2.5 Isolation of Genomic DNA

Take 1 ml of overnight bacterial culture grown in lb broth media and centrifuge at 12,000 rpm for 5 min. Resuspend the pellet in 100 μl of TE buffer and add 10 μl of lysozyme. Incubate at room temperature at 37 °C for 30 min. Add 20 μl of 10 % SDS and 15 μl of proteinase K. Incubate at 55 °C for 1 h or 60 °C for 30 min. Add 400 μl of 6 M ammonium acetate, pH-5.5. Vortex thoroughly and centrifuge at 12,000 rpm for 15 min. To the supernatant, add equal volume of isopropanol. Incubate at −20 °C for 30 min and centrifuge at 10,000 rpm for 10 min. Wash the pellet with 1 ml 70 % ethanol. Air dry the pellet and dissolve the pellet in 1× TE buffer.

1.2.6 Estimation of DNA: Spectrophotometric Determination

Analysis of UV absorption by the nucleotides provides a simple and accurate estimation of the concentration of nucleic acids in a sample. Purines and pyrimidines in nucleic acid show absorption maxima around 260 nm (e.g., dATP: 259 nm; dCTP: 272 nm; dTTP: 247 nm) if the DNA sample is pure without significant contamination from proteins or organic solvents. The ratio of OD_{260}/OD_{280} should be determined to assess the purity of the sample.

1.2.7 Polymerase Chain Reaction Application

To test the quality of the genomic preparation, primers were designed to amplify the triosephosphate isomerase gene of *L. delbrueckii* subsp. lactis using the TPI forward and reverse primers. Target DNA was cycled on an MJ Tetrad thermal cycler, Samples were analyzed on 2 % agarose gel containing ethidium bromide (Fig. 1.1).

1.2.8 Bioinformatics Analysis

SDSC BIOLOGY WORK BENCH is a tool kit which provides easy access for various bioinformatics tools to analyze the data. CLUSTAL W tool a product of SDSC was used to perform multiple sequence analysis of the TPI protein among different organisms. Phylogenetic study was made using the DENDROGRAM (Fig. 1.2). TEXSHADE and BOXSHADE were used to identify the common conserved regions among the sequences (Figs. 1.3, 1.4).

1.3 Results and Discussion

Different *Lactobacillus* strains will produce bacteriocins but the *Lactobacillus delbrueckii* strain is mostly used for the production of more amount of Bacteriocin as the TPI gene is only present in the *Lactobacillus delbrueckii* strain so the *Lactobacillus delbrueckii* is isolated from raw cow milk and the *Lactobacillus delbrueckii* strain were identified by several biochemical tests and the confirmatory test was mannitol test which shows no acid and no gas production which confirms the presence of the *Lactobacillus delbrueckii* and genomic DNA was isolated by proteinase K method and its TPI gene was amplified by using PCR. The Bacteriocin produced by the *Lactobacillus delbrueckii* is used as preserving agent in dairy and food industries and it used to prepare probitics to treat several diseases caused by bacteria. The Bacteriocin is used in the preparation of sporlac powder which is used to treat diarrhea. The Bacteriocin produced by the *Lactobacillus* strains is called lacticin which is mainly used in the food preserving and diary industries.

After amplification, the sequence analysis of TPI gene was done using the bioinformatics tools and the phylogenetic study was made using dendogram as it shows the evolutionary significance between the sequences. The dendogram infers that the query *Lactobacillus delbrueckii* species and the *Escherichia coli* falls under the same branch which shows that they share maximum similarity between the sequences whereas the Homo sapiens and Mus musculus falls under completely different branch which indicates evolutionary divergence. The conserved

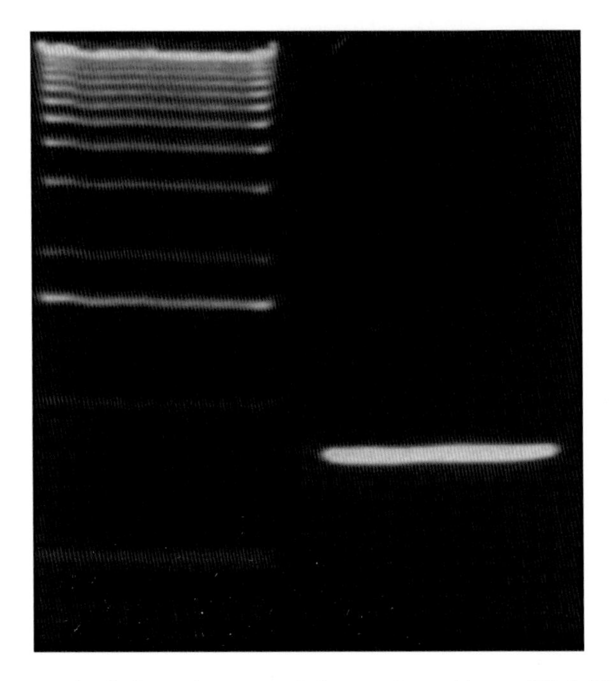

Fig. 1.1 Agarose gel of the polymerase chain reaction with amplified TPI gene from *Lactobacillus delbrueckii* subsp. lactis incorporating template DNA purified using our rapid procedure. Lanes: 1, 100 ng XVII500-bp PCR marker (10,000) 50 bp; 2 and 3, 4 ll of the corresponding PCR product directly using *L. delbrueckii*

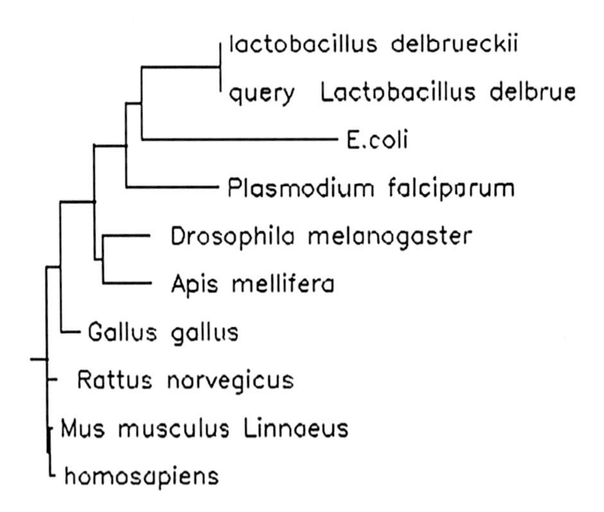

Fig. 1.2 The dendrogram indicates that query and *Lactobacillus delbrueckii* fall under common branch which shows that they share maximum similarity where as Homo sapiens falls under a completely different branch which indicates there evolutionary divergence

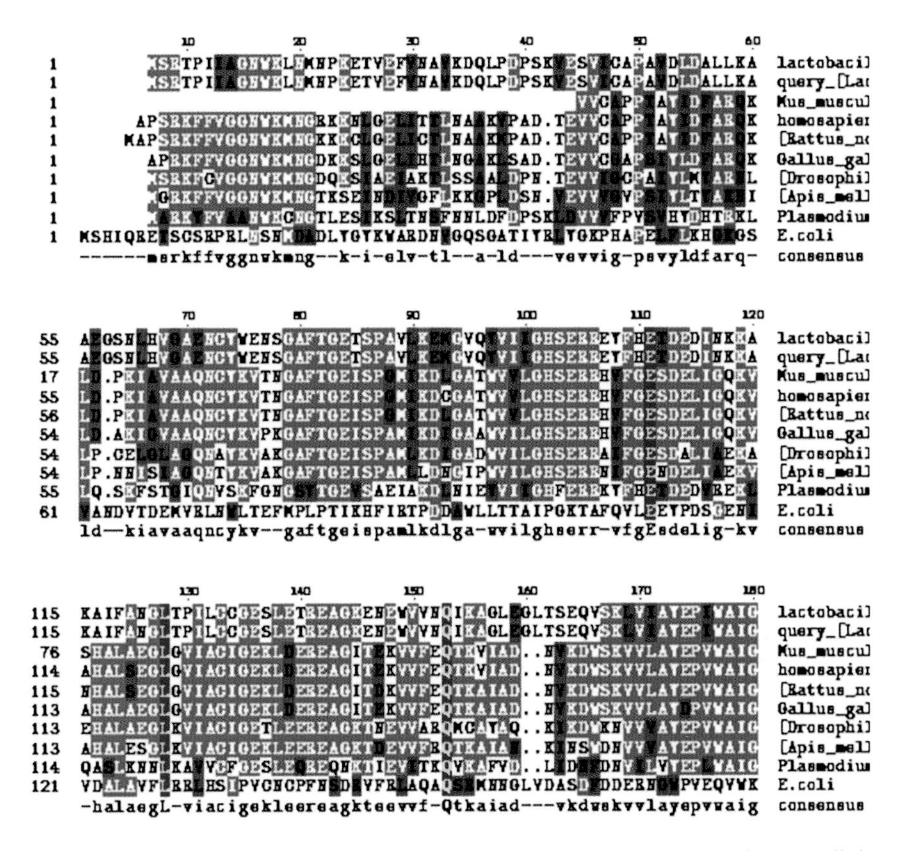

Fig. 1.3 This figure infers that most of the regions on the protein are conserved among all the organism

regions between the sequences were identified by using the text shade and box shade. In the text shade the blue color regions indicates the conserved regions, the pink color regions indicates the similar regions and the violet color regions indicates all the matching regions between the sequences. From the sequence analysis we can infer that the TPI gene sequence of *Lactobacillus delbrueckii* has close resemblance with the *E. coli* sequence. Hence we can conclude that the TPI gene plays a major key role in bacteriocin production and the project aimed to amplify the TPI gene and it was successfully amplified by using PCR and the amplified TPI gene will be further processed to produce bacteriocin which is used in the dairy, food industries and pharmaceutical companies.

```
lactobacillu■_d ------MSRTPIIAGNWRLNMNPRETVEFVNAVRDQLPDPSRVESVICAPAVDLDALLRA
query_[Lactobac ------MSRTPIIAGNWRLNMNPRETVEFVNAVRDQLPDPSRVESVICAPAVDLDALLRA
Mu■_mu■culu■_Li ---------------------------------------VVCAPPTAYIDFARQK
homo■apien■     -----APSRKFFVGGNWKMNGRKKNLGELITTLNAAKVPAD-TEVVCAPPTAYIDFARQK
[Rattu■_norvegi ----MAPSRKFFVGGNWKMNGRKKCLGELICTLNAAKMPAD-TEVVCAPPTAYIDFARQK
Gallu■_gallu■   ------APRKFFVGGNWKMNGDKKSLGELIHTLNGAKLSAD-TEVVCGAPSIYLDFARQK
[Dro■ophila_mel ------MSRKFCVGGNWKMNGDQKSIAEIAKTLSSAALDPN-TEVVIGCPAIYLMYARNL
[Api■_mellifera ------MGRKFFVGGNWKMNGTKSEINDIVGFLKKGPLDSN-VEVVVGVPSIYLTYARNI
Pla■modium_falc ------MARKYFVAANWKCNGTLESIKSLTNSFNNLDFDPSKLDVVVFPVSVHYDHTRKL
E.coli          MSHIQRETSCSRPRLNSNMDADLYGYKWARDNVGQSGATIYRLYGKPHAPELFLKHGKGS
consensus       ------m■rkffvggnwkmng--k-l-elv-tl--a-vd---vevvia-ptvyldfarq-

lactobacillu■_d AEGSNLHVGAENCYWENSGAFTGETSPAVLREMGVQYVIIGHSERREYFHETDEDINKKA
query_[Lactobac AEGSNLHVGAENCYWENSGAFTGETSPAVLREMGVQYVIIGHSERREYFHETDEDINKKA
Mu■_mu■culu■_Li LD-PKIAVAAQNCYKVTNGAFTGEISPGMIKDLGATWVVLGHSERRHVFGESDELIGQKV
homo■apien■     LD-PKIAVAAQNCYKVTNGAFTGEISPGMIKDCGATWVVLGHSERRHVFGESDELIGQKV
[Rattu■_norvegi LD-PKIAVAAQNCYKVTNGAFTGEISPGMIKDLGATWVVLGHSERRHVFGESDELIGQKV
Gallu■_gallu■   LD-AKIGVAAQNCYKVPKGAFTGEISPAMIKDIGAAWVILGHSERRHVFGESDELIGQKV
[Dro■ophila_mel LP-CELGLAGQNAYKVAKGAFTGEISPAMLKDIGADWVILGHSERRAIFGESDALIAEKA
[Api■_mellifera LP-NNISIAGQNTYKVAKGAFTGEISPAMLLDNGIPWVILGHSERRNIFGENDELIAEKV
Pla■modium_falc LQ-SKFSTGIQNVSKFGNGSYTGEVSAEIAKDLNIEYVIIGHFERRKYFHETDEDVKEKL
E.coli          VANDVTDEMVRLNWLTEFMPLPTIKHFIRTPDDAWLLTTAIPGKTAFQVLEEYPDSGENI
consensus       le--kiavaaqncykv--gaftgei■pamlkdmga-wvilgh■err-vfgE■delig-kv

lactobacillu■_d KAIFANGLTPILCCGESLETREAGKENEWVVNQIKAGLEGLTSEQVSKLVIAYEPIWAIG
query_[Lactobac KAIFANGLTPILCCGESLETREAGKENEWVVNQIKAGLEGLTSEQVSKLVIAYEPIWAIG
Mu■_mu■culu■_Li SHALAEGLGVIACIGEKLDEREAGITEKVVFEQTKVIAD--NVKDWSKVVLAYEPVWAIG
homo■apien■     AHALSEGLGVIACIGEKLDEREAGITEKVVFEQTKVIAD--NVKDWSKVVLAYEPVWAIG
[Rattu■_norvegi NHALSEGLGVIACIGEKLDEREAGITDKVVFEQTKAIAD--NVKDWSKVVLAYEPVWAIG
Gallu■_gallu■   AHALAEGLGVIACIGEKLDEREAGITEKVVFEQTKAIAD--NVKDWSKVVLAYDPVWAIG
[Dro■ophila_mel EHALAEGLKVIACIGETLEEREAGKTNEVVARQMCAYAQ--KIKDWKNVVVAYEPVWAIG
[Api■_mellifera AHALESGLKVIACIGEKLEEREAGKTDEVVFRQTKAIAN--KINSWDNVVVAYEPVWAIG
Pla■modium_falc QASLKNNLKAVVCFGESLEQREQNKTIEVITKQVKAFVD--LIDNFDNVILVYEPLWAIG
E.coli          VDALAVFLKKLHSIPVCNCPFNSDRVFRLAQAQSRMNNGLVDASDFDDERNGWPVEQVWK
consensus       -halaegL-viacigekleereagkteevvf-Qtkaiad---vkdw■kvvlayepvwaig

lactobacillu■_d TGKTASSDQAEEMCKTIRETVKDLYNEETAENVKIQYGGSVKPANVKELMAKPNIDGGLV
query_[Lactobac TGKTASSDQAEEMCKTIRETVKDLYNEETAENVKIQYGGSVKPANVKELMAKPNIDGGLV
Mu■_mu■culu■_Li TGKTATPQQAQEVHEKLRGWLKSNVNDGVAQSTRIIYGGSVTGATCKELASQPDVDGFLV
homo■apien■     TGKTATPQQAQEVHEKLRGWLKSNVSDAVAQSTRIIYGGSVTGATCKELASQPDVDGFLV
[Rattu■_norvegi TGKTATPQQAQEVHEKLRGWLKCNVSEEVAQCTRIIYGGSVTGATCKELASQPDVDGFLV
Gallu■_gallu■   TGKTATPQQAQEVHEKLRGWLKTHVSDAVAQSTRIIYGGSVTGGNCKELASQHDVDGFLV
[Dro■ophila_mel TGQTATPDQAQEVHAFLRQWLSDNISKEVSASLRIQYGGSVTAANAKELAKKPDIDGFLV
[Api■_mellifera TGKTATPQQAQEVHEKLRNWFSKNVNQTVAETVRIIYGGSVTAGNAKDLAKEKDIDGFLV
Pla■modium_falc TGKTATPEQAQLVHKEIRKIVKDTCGEKQANQIRILYGGSVNTENCSSLIQQEDIDGFLV
E.coli          EMHNLLPFSPDSVVTHGDFSLDNLIFDEGKLIGCIDVGRVGIADRYQDLAILWNCLGEFS
consensus       tgktatpqqaqevhekl r-wlk-nv-eevaq■vrIiyGg■vtganckeLa-qpdidGflv

lactobacillu■_d GGASLVPDSYLALVNYQD-------------
query_[Lactobac GGASLVPDSYLALVNYQD-------------
Mu■_mu■culu■_Li GGASLKPEFVDIINAKQ--------------
homo■apien■     GGASLKPEFVDIINAKQ--------------
[Rattu■_norvegi GGASLKPEFVDIINAKQ--------------
Gallu■_gallu■   GGASLKPEFVDIINAKH--------------
[Dro■ophila_mel GGASLKPEFVDIINAKQ--------------
[Api■_mellifera GGASLKPDFVQIVNAKQ--------------
Pla■modium_falc GNASLKESFVDIIKSAM--------------
E.coli          PSLQKRLFQKYGIDNPDMNKLQFHLMLDEFF
consensus       gga■lkpefvdiinakq--------------
```

Fig. 1.4 This figure infers that most of the regions on the protein are conserved among all the organism

References

1. Mullis K (1993) Nobel lecture. World Scientific Publishing, Singapore
2. Bartlett JM, Stirling D (2003) A short history of the polymerase chain reaction. Methods Mol Biol 226:3–6
3. Farkas-Himsley H, Yu H (1984) Purified colicin as cytotoxic agent of neoplasia: comparative study with crude colicin. Cytobios 42(167–168):193–207
4. Baumal R, Musclow E, Farkas-Himsley H, Marks A (1982) Variants of an interspecies hybridoma with altered tumorigenicity and protective ability against mouse myeloma tumors. Cancer Res 42(5):1904–1908

5. Cotter PD, Hill C, Ross RP (2006) Nat Rev Microbiol 4(2) doi:10.1038/nrmicro1273c2. (http://www.nature.com/nrmicro/journal/v4/n2/full/nrmicro1273-c2.html is author reply to comment on article: Cotter PD, Hill C, Ross RP (2005) Bacteriocins: developing innate immunity for food. Nat Rev Microbiol 3 (10):777–788. doi:10.1038/nrmicro1273. PMID: 16205711)
6. Del Rio B, Binetti AG, Martın MC, Fernandez M, Magadan AH, Alvarez MA (2007) Multiplex PCR for the detection and identification of dairy bacteriophages in milk. Food Microbiol 24:75–81
7. Dalhus B, Fimland G, Nissen-Meyer J, Johnsen L (2005) Pediocin-like antimicrobial peptides (class IIa bacteriocins) and their immunity proteins: biosynthesis, structure, and mode of action. J Pept Sci 11(11):688–696. doi:10.1002/psc.699 (PMID: 16059970)
8. Cascales E, Buchanan SK, Duché D et al (2007) Colicin biology. Microbiol Mol Biol Rev 71(1):158–229. doi:10.1128/MMBR.00036-06 (PMID: 17347522)
9. Cruz-Chamorro L, Puertollano MA, Puertollano E, de Cienfuegos GA, de Pablo MA (2006) In vitro biological activities of magainin alone or in combination with nisin. Peptides 27(6):1201–1209. doi:10.1016/j.peptides.2005.11.008 (PMID: 16356589)
10. Cheng S, Fockler C, Barnes WM, Higuchi R (1994) Effective amplification of long targets from cloned inserts and human genomic DNA. Proc Nat Acad Sci 91(12):5695–5699. doi:10.1073/pnas.91.12.5695 (PMID: 8202550)
11. Gratia JP (2000) André Gratia: a forerunner in microbial and viral genetics. Genetics 156(2):471–476. (PMID: 11014798, PMC: 1461273; http://www.genetics.org/cgi/pmidlookup?view=long&pmid=11014798)
12. Sambrook J, Russel DW (2001) Molecular cloning: a laboratory manual, 3rd edn. Cold Spring Harbor Laboratory Press, Cold Spring Harbor

Chapter 2
Lactobacillus Model Moiety a New Era Dosage Form as Nutraceuticals and Therapeutic Mediator

Abhinandan R. Patil, Sunita S. Shinde, Pratik S. Kakade and John I. D'souza

Abstract In this era Bacteria are mostly considered as pathogenic. Bacteria though are pathogenic but still a few are friendly and essential for human growth and immunity; called as Probiotics. *Lactobacillus* model moiety is one of such probiotic bacterium which when introduced in sufficient colony serves as new prophylaxis and curing agent. This will increase the innate immunity of humans. Indian market lacks the quality uni-strain product of the probiotics as Nutraceuticals and as Health enhancing drug product. Global market is full of multi-strain microbes dosage form. The best mode to utilize Nutraceutical aspect of probiotics is to convert in dry form. These purified colony; later converted into the solid dry form by spray dry (JISL mini-spray drier) technique. Starch, lactose were used as thermo-protective agent while heat drying technique. Anti-microbial screenings carried out along with In vitro cytotoxicity studies. In vitro cytotoxicity screening evaluated that *Lactobacillus* powder showings anti-cancer activity nearly same as standard drug with no side effects. Dry powder increased the shelf life of the microbes that resulted in maintenance of viability and activity.

Keywords *Lactobacillus* model moiety · Nutraceuticals · Probiotics · Pathogenic · Uni-strain

2.1 Introduction

There are more bacteria in the world today than all the humans ever born. Most of them considered as pathogen. As every coin had, two sides besides mostly bacteria are pathogen but still few are friendly essential for human growth and immunity.

A. R. Patil (✉) · S. S. Shinde · P. S. Kakade · J. I. D'souza
Department of Pharmaceutics, Tatyasaheb Kore College of Pharmacy, Warananagar, Kolhapur 416113, Maharashtra, India
e-mail: arpatil.tkcp@gmail.com

A. Kumar, *Biotechnology and Bioforensics*, Forensic and Medical Bioinformatics,
DOI: 10.1007/978-981-287-050-6_2,

Such friendly bacteria are called probiotics. The following are microorganisms considered to be human probiotics: *Lactobacillus species:* ***Lactobacillus model moiety:*** *L. amylovorus* etc., *Bifidobacterium* species: *B. adolescentis*, etc., other lactic acid bacteria: *Enterococcus faecium, Lactococcus lactis, Leuconstoc mesenteroides, Pediococcus acidilactici, Streptococcus thermophilus* and Nonlactic acid bacteria: *Bacillus subtilis* etc.

Lactobacillus organisms are normal inhabitants of the human intestine and vagina. *Lactobacilli* are gram-positive facultative anaerobes; non-spore forming; and non-flagellated, rod or coccobacilli. To date, some 56 species of *Lactobacillus* have been identified.

- *Lactobacillus model moiety* is the most commonly known probiotic bacterium. It is found primarily in the small intestine where it produces natural antibiotics called "lactocidin" and "acidophilin". These increase immune resistance against such harmful bacteria and fungi as *Candida albicans, Salmonella, E. coli*, and *Staphylococcus aureus*.
- *Lactobacillus model moiety* implants itself the intestinal walls, as well as on the lining of the vagina, cervix, and urethra, thereby preventing other organisms from multiplying to the extent that they can cause infections. For years, it was assumed that it was the most beneficial form of the "good" bacteria; but recent research has revealed that *L. rhamnosus* may be of more importance.
- *Lactobacillus model moiety* helps control intestinal infections, thus reducing the potential of diarrhea and other infections or diseases. It also inhibits some types of cancer and helps control serum cholesterol levels. However, reaching the intestines is the problem because the *Lactobacillus model moiety* found in most commercial yogurts cannot live with stomach acids and bile.

Mostly present recently in dairy product but very rarely used in form of dosage form for treatment and as prophylaxis of diseases. To sub-culture the friendly microbes of *lactobacilli* in MRS (de Man, Rogosa and Sharpe) media is unique feature in form of Colony forming unit (CFU). To produce agent alternative to Antibiotics as Probiotics produces natural antibiotics called "lactocidin" and "acidophilin" is basic theme of formulation generation that will contribute to generate Innate immunity. Thus this enhances friendly micro flora in Gastrointestinal tract (GI) to maintain proper bowl movement. Spray drying is one of the oldest methods of encapsulation used initially for flavour capture. It is a single step continuous processing operation. The process can produce purest and finest powders with high sterility reducing the post unit operation like grinding and conditioning. Spray dried powder particles are relatively small and uniform in size and shape. Spray drying allows a uniform dispersion of powder particle by diluting the bioactive core when a low amount is required. Sulphorhodamine (SRB) assay give general idea to screen anti-cancer activity.

2.2 Materials and Methods

2.2.1 Preparation of Bacterial Suspension

Lactobacilli Species (*Lactobacillus model moiety*) were obtained from the "Warana Diary" (Warananagar) M.S; for long-term maintenance, this organism was stored as glass bead cultures in freezer at −20 °C. Man, Rogosa, Sharpe—MRS broth Obtained from the "Siffin Pharma" (Germany) (Fig. 2.1).

Latter these microbes were harvested into the natural media like pasteurised milk of cow and buffalos.

2.3 Methodology

The exact four procedure applied to generate the formulation as follows.

2.3.1 Part A

The milk of cow/buffalo were pasteurised before use to nil the other microbes if present in the milk. The pasteurization and tyndallisation gave the platform to the new inoculated microbes to get flourish into the milk after cooling of this milk. Tyndallisation was carried out in beaker for 3 days and then microbes were inoculated to get semi-solids beads (Fig. 2.2).

2.3.2 Part B: Homogenization and Spray Dry

This semi-solid mass obtained by 24 h incubation in oven were broken down/converted into liquid state by Homogenization. 2,300 rpm (rotation per min) used to break the bead/semi solid mass formation. The spray-drying process of *Lactobacilli* Species in the various media was undertaken in a laboratory scale spray dryer (Jisl mini-spray dry). The excipients used for the spray dry were lactose and starch solution. This combination of lactose and starch (2:1) acts as the thermoprotective agent to prevent the mortality of the microbes or decrease in the cell count while at the spray dry procedure due to the heavy inlet temperature. The viability at different combination were carried out for lactose and starch solution as 1:1, 1:2, 2:1 etc. to check for good results (Fig. 2.3).

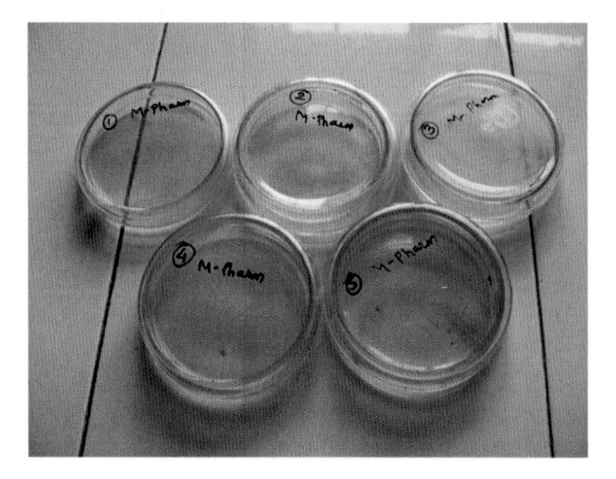

Fig. 2.1 Multiple plates

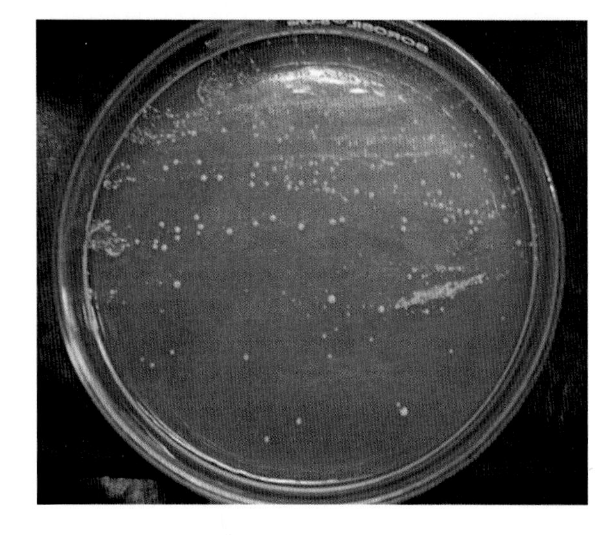

Fig. 2.2 Plate of MRS

2.3.3 Part C: In Vitro Cytotoxicity Studies: SRB Assay

Cell line: Colon cancer HT-29 procured from (N.C.C.S) National Centre for Cell Science, Pune. Standard anticancer drug Capacitabin obtained as gift sample from the Atmatara foundation and research unit (Kolhapur). Stock solution of *Lactobacillus model moiety* powder of 1 mM in 0.25 % Dimethyl sulphoxide (DMSO) was prepared and further dilution done in 10, 50 and 100 μM with phosphate buffer saline (PBS). Stock solution of Capecitabine of 1 mM in distilled water was prepared and further dilution done in 10, 50 and 100 μM with phosphate buffer saline (PBS).

Fig. 2.3 Spray dry JISL

2.3.4 Part D: Anti-microbial Screening (Well Assay)

MRS agar plates were prepared; that surface of the MRS agar plates was inoculated with the swab containing 24 h culture *V. cholera* and for *S. dysentriae*. Wells were punched with the gel puncher 80 mg *Lactobacillus model moiety* powder (L.A.P.) were added to the wells punched in the centre of the plates. Plates were incubated for 24 h at 37 °C. *S. dysentriae* were also incubated on the MRS agar plates, separately and wells were punched. Wells, with a diameter of 5 mm, then cut in the agar using sterile gel puncher.

2.4 Results and Discussion

The results for the **Part A** work were in form of the CFU (colony forming units) count as,

Fig. 2.4 CFU count in semi solids mass form in beaker

Table 2.1 Batch dilution at 10^6 CFU/ml by plate count for the cow milk

Sr. no.	1:1	1:2	2:1
1	12	15	35
2	12	14	34
3	12	14	34

- Count in Milch animals

  (a) Cow = 10^6 CFU/ml (colony forming units).
  (b) Buffalos = 10^7 CFU/ml (colony forming units).

This difference in the cell count was due to the difference in the fat, nutrients avail in both animals milk (Fig. 2.4).

The results for the **Part B and Part C** work were obtained in form liquid generated at the rotation per min (rpm) via Homogenization carried out from 1,000 to 2,300 rpm. The results seen in case at 2,300 rpm for 20 min were optimal. This solution obtained by this above procedural was acceptable for the next spray dry mythology. The large size particle generally clogged the spray dry at the gun nozzle but size reduction at 2,300 rpm for 20 min prevented the above discussed problem.

The spray drying carried out by taking the combination of the Lactose and starch in different ratios and the cell count obtained by serial dilution for the cow milk is given in Table 2.1.

The results were optimal for the cow milk as by spray dry. The 2:1 ratio of (lactose and starch) gave good cell count as 35×10^6 CFU/ml (average).

Similarly, the spray drying carried out by taking the combination of the Lactose and starch in different ratios and the cell count obtained by serial dilution for the Buffalo milk is given in Table 2.2.

Table 2.2 Batch dilution at 10^6 CFU/ml by plate count for the buffalos milk

Sr. no.	1:1	1:2	2:1
1	12	14	21
2	12	15	21
3	12	14	22

Table 2.3 Other thermo protective agent used and there effects

Nature/excipients	Lactose	Starch	Lactose + Starch
Powder nature	Amorphous/nonhygroscopic	Hygroscopic	Amorphous/hygroscopic
Cell count	Less	Less	Good

The results were optimal for the Buffalos milk as by spray dry. The 2:1 ratio of (lactose and starch) gave good cell count as 21×10^6 CFU/ml (average).

2.4.1 Other Thermo Protective Agent Used and There Effects Observed by Spray Dry

The spray dry results obtained in case of the Lactose alone or starch alone is not so good than lactose and starch used in combination at (2:1) ratios is given in Table 2.3.

2.4.2 Survival Rate During Spray Drying

The powder were collected and packed in the air tight container. The mortality rate is less by the utilization of the lactose and the starch in the composition (2:1) kept in dark air tight container at room temperature than keeping in open containers.

The power viability or cell count has been checked by simple serial dilution technique for both cow and buffalos spray dried powder as:

For 1 g in 10 ml = 10×10^6 CFU/ml *approx in case of cow;
1 g in 10 ml = 70×10^6 CFU/ml *approx in case of buffalo spray dried powder.

2.4.3 Part D: In Vitro Cell Cytotoxicity Studies

(Table 2.4) (Fig. 2.5).

Table 2.4 SRB assay

Concentration (μM/ml)	Percentage cell cytotoxicity [a](L.A.P.)/test	Percentage cell cytotoxicity [b]std. drug
10	48.87 ± 0.9536	50.23 ± 0.5752
50	68.56 ± 0.6894	69.85 ± 0.5275
100	97.83 ± 0.3768	86.53 ± 0.3217
IC_{50} value	11	9

[a] (L.A.P.) = *Lactobacillus model moiety*
[b] Std. drug = Capacitabin/capa.

Fig. 2.5 SRB assay

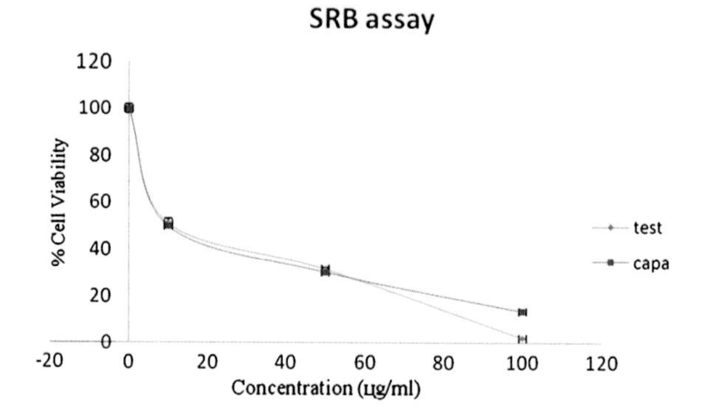

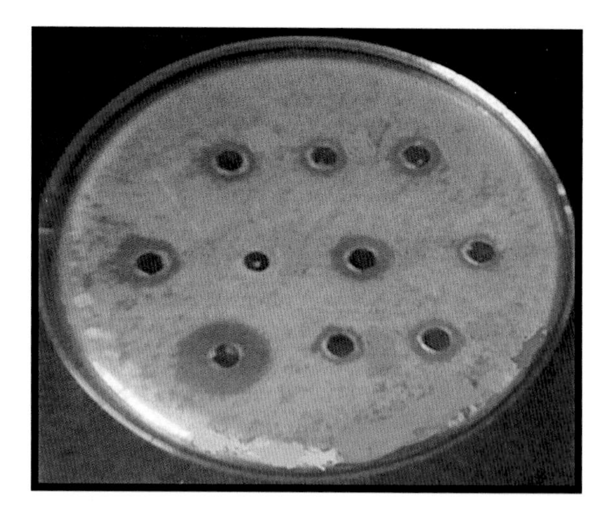

Fig. 2.6 Zone of inhibition on *Vibrio cholera*

2.4.4 Part E

The inhibitory activity against *V. cholera* and *S. dysentriae* was seen in form of clearance obtained with each organism as (Figs. 2.6, 2.7):

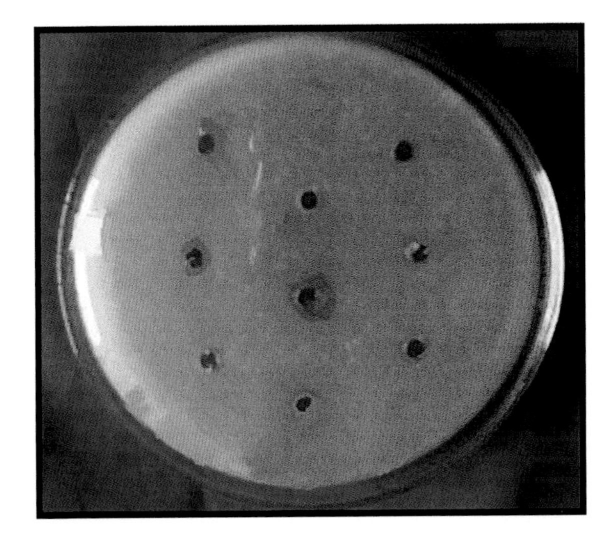

Fig. 2.7 Zone of inhibition of *Shigela dysentriae*

Table 2.5 Zone of inhibition study

Sr. no.	Name of organism	Zone of inhibition (mm)
1	*Vibrio cholera*	9
2	*Shigella dysentriae*	7

2.4.5 Organisms Zone of Clearance

See Table 2.5.

2.5 Conclusion

Dry powder increased the shelf life of the microbes and the formulation can be used any time by mere opening the pack kept at the airtight container. Lactose and starch given optimized thermo protection effect that prevented the cell death by heat in the spray dry temperature of inlet heat of 100–110 °C.

Revalidation of the powder carried out by serial dilution method in which the spray dry powder is taken in the saline water and is added in 10 ml water, and then shaken well. From that 10 ml solution 1 ml of the sample is transfer in next 10 ml test tube and shaken and diluted to 10 times. This procedure repeated till 10^6 times. Latter plates of MRS media were prepared to check cell count, which was found optimal in case of the buffalos spray dry powder than cow's powder. Best Nutraceutical agent was been accessed by antibiotic effects seen successfully in case of *V. cholera* and *S. dysentriae* zone of inhibition study. Better choice therapeutic mediators in cancer study were verified by SRB test.

Bibliography

1. Vinderola G, Capellini B, Villarreal F, Suaŕez V, Quiberoni A, Reinheimer J (2008) Usefulness of a set of simple in vitro tests for the screening and identification of probiotic candidate strains for dairy use. J Food Sci Technol 41:1678–1688
2. Iannitti T, Palmieri B (2010) Therapeutical use of probiotic formulations in clinical practice. Clin Nutr 29:701–725
3. Madhwani T, McBain A (2011) Bacteriological effects of *a Lactobacillus reuteri* probiotic on in vitro oral biofilms. Arch Oral Biol 56:1264–1273
4. Preidis G, Versalovic J (2009) Targeting the human microbiome with antibiotics probiotics, and prebiotics: gastroenterology enters the metagenomics era. Gastroenterology 136:2015–2031
5. Kaur I, Chopra K, Saini A (2002) Probiotics: potential pharmaceutical applications. Eur J Pharm Sci 15:1–9
6. Gupta V, Garg R (2009) Probiotics. Indian J Med Microbiol 27(3):202–209
7. Bernstein C et al (2009) Inflammatory bowel disease: a global perspective. World Gastroenterology Organisation Global Guidelines, USA, pp 1–24
8. Macfarlane S, Furrie E, Kennedy A, Cummings J, Macfarlane G (2005) Mucosal bacteria in ulcerative colitis. Br J Nutr 93:67–72
9. Hatakka K, Holma R, El-Nezami H, Suomalainen T, Kuisma M, Saxelin M, Poussa T, Mykkänen H, Korpela R (2008) The influence of *lactobacillus rhamnosus LC705* together with *Propionibacterium freudenreichii ssp. shermanii JS* on potentially carcinogenic bacterial activity in human colon. Int J Food Microbiol 128:406–410
10. Roller M, Clune Y, Collins K, Rechkemzer G, Watzl B (2007) Consumption of prebiotic inulin enriched with oligofructose in combination with the probiotics *Lactobacillus rhamnosus* and *Bifidobacterium lactis* has minor effects on selected immune parameters in polypectomised and colon cancer patients. Br J Nutr 97:676–684
11. Rafter J, Bennett M, Caderni G, Clune Y, Hughes R, Karlsson C, Klinder A, O'Riordan M, O'Sullivan C, Pool-Zobel B, Rechkemmer G, Roller M, Rowland I, Salvadori M, Thijs H, Van Loo J, Watzl B, Collins K (2007) Dietary synbiotics reduce cancer risk factors in polypectomized and colon cancer patients. Am J Clin Nutr 85:488–496
12. Liong M (2008) Safety of probiotics: translocation and infection. Nutr Rev 66(4):192–202
13. Lewandowicz G, Harding B, Harkness W, Hayward R, Thomas DG, Darling L (2000) Chemosensitivity in childhood brain tumours in vitro: evidence of differential sensitivity to lomustine (CCNU) and vincristine. Cancer 36:1955–1964
14. Tanaka T (2009) Colorectal carcinogenesis: review of human and experimental animal studies. J Carcinog 8:1–19
15. Tortora G, Grabowski S (2003) Principles of anatomy and physiology, 10th edn. Wiley, New Jersey, pp 853–855, pp 891–893
16. Levi E, Misra S, Du J, Patel B, Majumdar P (2009) Combination of aging and dimethylhydrazine treatment causes an increase in cancer—stem cell population of rat colonic crypts. Biochem Biophys Res Commun 385:430–433
17. Ahmad S, Anjum F, Huma N, Sameen A, Zahoor T (2013) Composition and physico-chemical characteristics of buffalo milk with particular emphasis on lipids, proteins, minerals, enzymes and vitamins. J Anim Plant Sci 23:62–74
18. Shah NP, Ravula RR (2002) Microencapsulation of probiotic bacteria and their survival in frozen fermented dairy desserts. Aust J Dairy Technol 55(3):139–144
19. Adhikari K, Mustapha A, Grun IU (2003) Survival and metabolic activity of microencapsulated *Bifidobacterium longum* in stirred yoghurt. J Food Sci 68(1):275–280
20. Albertini B, Passerini N, Pattarino F, Rodriguez L (2008) New spray-congealing atomizer for the microencapsulation of highly concentrated solid and liquid substances. Eur J Pharm Biopharm Sci 69:348–357

21. Albertini B, Passerini N, Di-Sabatino M, Vitali B, Brigidi P, Rodriguez L (2009) Polymer–lipid based mucoadhesive microspheres prepared by spray congealing for the vaginal delivery of Econazole nitrate. Eur J Pharm Sci 36:591–601
22. Albertini B, Vitali B, Passerini N, Cruciani F, Di-Sabatino M, Rodriguez L, Brigidi P (2010) Development of microparticulate systems for intestinal delivery of *Lactobacillus acidophilus* and *Bifidbacterium lactis*. Eur J Pharm Sci 40:359–366
23. Rault A, Bouix M, Béal C (2010) Cryotolerance of *Lactobacillus delbrueckii* subsp. *bulgaricus CFL1* is influenced by the physiological state during fermentation. Int Dairy J 20:792–799
24. Wargovich MJ et al (2000) Efficacy of potential chemopreventive agent on rat colon aberrant crypt formation and progression. Carcinogenesis 21(6):1149–1155
25. Fasoli S et al (2003) Bacterial composition of commercial probiotic products as evaluated by PCR-DGGE analysis. Int J Food Microbiol 82:59–70
26. Ananta E et al (2005) Cellular injuries and storage stability of spray-dried *Lactobacillus rhamnosus GG*. Int Dairy J 15:399–409
27. Semyonov D et al (2010) Microencapsulation of *Lactobacillus paracasei* by spray freeze drying. J Food Res Int 43:193–202
28. Albertini B et al (2010) Development of microparticulate systems for intestinal delivery of *Lactobacillus acidophilus* and *Bifidobacterium lactis*. Eur J Pharm Sci 40:359–366

Chapter 3
Decolorization and Biosorption of Dyes Using *Aspergillus* Sp.

Sahithi Dhatrika and T. Pravin Reddy

Abstract Release of toxic and recalcitrant chemicals including synthetic dyes from industries profoundly affects soil fertility and aquatic life. Use of physical and chemical methods for removal of dyes creates disposable problem of remaining dye sludge, whereas biotechnological approach provides viable, less sludge and eco-friendly method. In the present study, a fungus *Aspergillus* sp. isolated from the soil was employed for decolorization and degradation of five dyes. At first the biodegradation of these dyes was first studied in potato dextrose broth and maximum decolorization of 80.86, 40.05, 92.44, 38.12 and 95 % of decolorization was observed for Methylene blue, Bromophenol blue, Congo red, Malachite green and Rose Bengal respectively. As the maximum decolorization was observed with Congo red and Rose Bengal, the effect of different parameters like carbon source, nitrogen source, pH and temperature was also studied in these dyes. Maximum decolorization of (91 and 52 %) was achieved when supplemented with carbon source, (87 and 46 %) with nitrogen source, (86 and 51 %) at pH at 4–5, (90 and 41 %) at 25–30 °C for Congo red and rose Bengal respectively. The degradation of dyes was observed by the change in original color and visual disappearance of color from the fungus-treated cultures. Degradation/decolorization of dyes was also observed as accumulation of dyes by the fungal mycelium, and it was confirmed by the presence of colored fungal mycelium in fungus-treated cultures. Biosorption studies were also done by using different concentrations of dyes and dead and inactive mycelial mat of *Aspergillus* sp. This indicated that binding of the dye to dead mycelial mat occurred rapidly with more than 50 % within 1 day and the dye absorbed to the mat increased with increase in the dye concentration, a maximum of 93 and 90 % decolorization was attained at 5 mg Congo red and rose Bengal concentrations with same mycelia mat and maximum

---

S. Dhatrika (✉)
Indian Institute of Chemical Technology (IICT), Tarnaka, Andhra Pradesh, India
e-mail: sahithidhatrika@gmail.com

T. Pravin Reddy
Loyola Degree and PG College, Old Alwal, Andhra Pradesh, India

A. Kumar, *Biotechnology and Bioforensics*, Forensic and Medical Bioinformatics,
DOI: 10.1007/978-981-287-050-6_3, © The Author(s) 2015

of 92 and 88 % decolorization was attained at 750 mg of dead mycelial mat concentration with same dye concentration. Thus the decolorization achieved by metabolically inactive mycelium was equal to that attained by the live mycelium and *Aspergillus* sp. showed maximum decolorization with Congo red and Rose Bengal dyes. Further this can be commercially used to treat industrial effluents.

Keywords Biosorption · *Aspergillus* species · Decolorization

3.1 Introduction

Over the last few decades, the increasing use of synthetic dyes is alarming and their discharge as textile waste may cause substantial ecological damage. The effluents from textile, leather, food processing, dyeing, cosmetics, paper, and dye manufacturing industries are important sources of dye pollution. The source of such pollution lies in the rapid increase in the use of synthetic dyes. More than 10,000 chemically different dyes are being manufactured. The world dyestuff and dye intermediates production is estimated to be around 7–108 kg per annum [1, 2]. The effluents from industries, which include both dye manufacturing and dye application, are highly colored. Strong color of the textile waste water is the most serious problem of the textile waste effluent. The disposal of these wastes into receiving water causes damage to the environment. Dyes may significantly affect photosynthetic activity in aquatic habit because of reduced light penetration and may also toxic to some aquatic life due to the presence of aromatics, metals, chlorides and other toxic compounds [3].

The textile dyes are highly reactive and the most commonly used are azo dyes. Azo dyes, which are aromatic compounds with one or more (–N=N–) groups, are the most important and largest class of synthetic dyes used in commercial applications [4]. These synthetic dyes in wastewater cannot be efficiently decolorized by traditional methods such as Physico-chemical methods like Adsorption, Sedimentation, Coagulation, Membrane filtration, Ozonisation, Chlorination etc. that are quite expensive, low efficiency and inapplicability to a wide variety of dyes and in addition these methods do not generally degrade the pollutant, thereby causing an accumulation of the dye as sludge creating a disposal problem. These technologies have been reviewed by Robinson [5], and therefore special attention is given to biological processes because they are with low cost, high efficiency and environmentally friendly. It has been demonstrated that mixed bacterial cultures are capable of decolorizing textile dye solutions. Nevertheless, several studies show that little biodegradation actually occurs and that the primary mechanism is adsorption to the microbial biomass [6]. In recent years, there has been a alternative research on fungal decolorization of dye present in wastewaters, and it is turning into a promising alternative to replace or supplement for present treatment processes [7]. Biosorption is also becoming a promising alternative to replace or supplement the present dye removal processes from dye wastewater.

3.2 Materials and Methods

Organism Used: A Soil isolate of *Aspergillus* sp., was used for the decolorization studies of dyes. The fungi strain was maintained on Potato dextrose agar slants at 30 °C.

Dyes Used: Congo red dye, Bromophenol blue dye, Malachite green dye, Methylene Blue dye and Rose Bengal dye were obtained from Ranbaxy Laboratories Limited, S.A.S. Nagar, Punjab.

Media Used: Basal medium, Czapek Dox Broth, Potato dextrose agar medium were obtained from Hi media Laboratories Private Limited, Mumbai.

Biodegradation of Different Dyes Using Isolated Fungi: A fungal strain isolated from the dye containing effluents was identified and pure cultures of were prepared and the influence of isolated on the removal of several dyes was determined by taking 100 ml of each batch culture medium into 250 ml Erlenmeyer flasks and was supplemented with Congo red, Bromophenol blue, Malachite green, Methylene Blue, and Rose Bengal separately with a concentration of 1, 2 and 5 mg to all the flasks and autoclaved, then two agar discs of 6 mm containing the isolated fungal strain were inoculated and incubated in an orbital shaker. Non-inoculated culture medium was used as control. Samples were drawn at regular interval 1–5 days and absorbance was measured. Mycelia were collected by filtration and the supernatant was analyzed for dye decolorization.

Dye disappearance was determined spectrophotometrically by monitoring the absorbance at or near the wave length maximum for each dye (500 nm for Congo red, 590 nm for Bromophenol blue dye, 621 nm for Malachite green dye, 670 nm for Methylene Blue dye, 525 nm for Rose Bengal dye). Experiment was performed in triplicates and mean average values were taken. The decolorization efficiency (DE) expressed in % of the initial dye concentration was calculated as follows—$DE = 100 \times (OD_i - OD_f)/OD_i$, where OD_i is the absorbance value of the initial dye concentration and OD_t is the absorbance value of the final dye concentration at time t.

3.3 Biosorption Experiment

For biosorption experiment, mycelia mat of isolated fungi was obtained by growing the isolated culture of fungus on Czapeck Dox medium in the absence of dye for 3 days. This culture was autoclaved at 121 °C for 15 min and filtered to collect the dead mycelia biomass. The dead and inactive mycelium was washed in distilled water, air dried on filter paper for 2–3 h and weighed. Two different biosorption experiments were designed

1. By transferring equal amounts (500 mg) of dead mycelia into 100 ml conical flasks at different dye concentrations at a range of 40–200 ppm.
2. By distributing uniform initial dye concentration of 200 ppm in all flasks but transferring different amounts of fungal biomass in a range of 350–800 mg.

Aliquots of supernatant free of mycelia were taken at different time intervals of incubation and approximately diluted to measure the absorbance at 559 nm. At the end of the experiment, dry weight of biomass was obtained by drying the filtered mycelium at 50 °C for 3 h. % Decolorization is expressed on the basis of dry weight. % Decolorization was calculated using the formula—$DE\% = 100 \times (OD_i - OD_f)/OD_i$, where OD_i is the absorbance value of the initial dye concentration and OD_t is the absorbance value of the final dye concentration at time t.

3.4 Results and Discussion

A soil isolate of fungi which has showed the ability to decolorize the rose Bengal present in the chloramphenicol Rose Bengal agar plate was used in the present study. Pure culture of the fungi was isolated and maintained on potato dextrose agar slants. The pure cultures were further analyzed for identification of fungi using slide culture technique and SEM based on the morphological characteristics. As shown in Table 3.1 the isolated fungi showed less decolorization towards Bromophenol blue, Malachite green, Methylene Blue whereas with Congo red and Rose Bengal dyes, the isolated fungi showed maximum decolorization. Thus Congo red dye and rose Bengal dye were used to determine the factors like carbon source (Glucose and fructose), nitrogen source (Yeast extract and Ammonium nitrate), pH (4, 5, 6, 8 and 9) and temperature (25, 30 and 40 °C) affecting the decolorization by fungus.

Influence of Adding Carbon and Nitrogen Source on the Rate of Decolorization of Congo Red: The effect of carbon sources (glucose and fructose) and Nitrogen (Yeast Extract and NH_4NO_3) on the rate of decolorization by *Aspergillus* sp. was studied and from the graph, it is inferred that both carbon sources showed similar percentage of decolorization of 91 % of the Congo red. As the incubation period has increased dye degradation increased from 43 to 90 % in shaking condition and 55 to 90 % under static conditions.

Table 3.1 Biodegradation of different dyes with different concentrations

Days	Methylene blue			Bromo phenol blue			Congo red			Malachite green			Rose bengal		
	MB 1 mg	MB 2 mg	MB 3 mg	BPB 1 mg	BPB 2 mg	BPB 3 mg	CR 1 mg	CR 2 mg	CR 3 mg	MG 1 mg	MG 2 mg	MG 3 mg	RB 1 mg	RB 2 mg	RB 3 mg
0	0	0	0	0	0	0	0	0	0	0	0	0	0	0	0
1	10.08	8.3	7.16	30.33	30.37	30.03	54.36	55.65	86.44	13.4	4.06	21.21	80.2	82.2	92.76
2	19.83	19.51	17.56	89.89	93.69	35.42	63.53	77.58	89.38	21.6	11.08	25.78	81.57	91.31	94.89
3	95.92	61.96	60.78	91.39	94.93	43.5	65.92	77.92	90.09	29.1	26.66	31.57	83.21	91.65	94.18
4	95.99	73.63	69.69	94.76	95.66	43.58	68.03	78.6	90.84	32.5	32.85	32.76	86.21	92.56	95.12
5	96.1	83.91	80.86	95.02	93.13	40.05	71.27	79.22	92.44	47.0	40.06	38.12	88.48	92.98	95.8

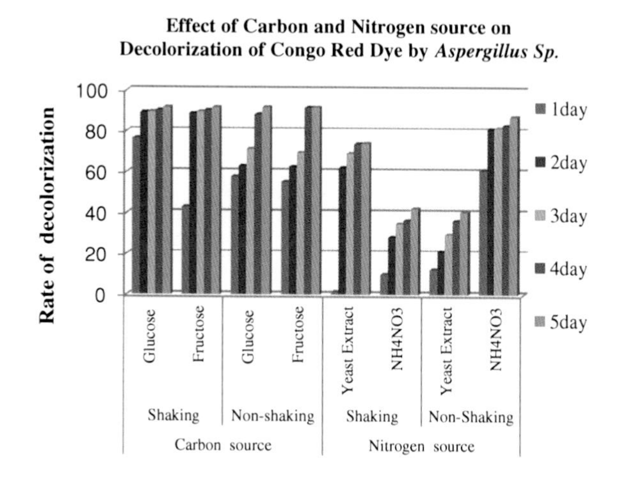

Among the Nitrogen sources, NH_4NO_3 was the best for Congo red decolorization which increased the decolorization from 60 to 87 % at the end of 5th day under non shaking conditions and under shaking conditions yeast extract was the best for Congo red decolorization which increased the decolorization from 1 to 74 % at the end of 5th day.

Influence of Adding Carbon and Nitrogen Source on the Rate of Decolorization of Rose Bengal: The effect of carbon sources (glucose and fructose) and Nitrogen (Yeast Extract and NH_4NO_3) on the rate of decolorization by *Aspergillus* sp. was studied. As shown in the graph these two carbon sources are able to degrade more in static conditions as compared to shaking conditions. The percentage of degradation is 52 and 49 % in static conditions whereas in shaking conditions it showed only 37 and 25 %. Among the Nitrogen sources, ammonium nitrate was the best for Rose bengal decolorization which increased the decolorization from 44 to 46 % at the end of 5th day while yeast extract was little less in dye decolorization as it decolorized rose bengal from only 27 to 37 % at the end of 5th day under non shaking conditions and under shaking conditions yeast extract was the best for rose bengal decolorization which increased the decolorization from 18 to 30 % at the end of 5th day while ammonium nitrate was poor in dye decolorization as it decolorized Congo red from only 16 to 22 % at the end of 5th day.

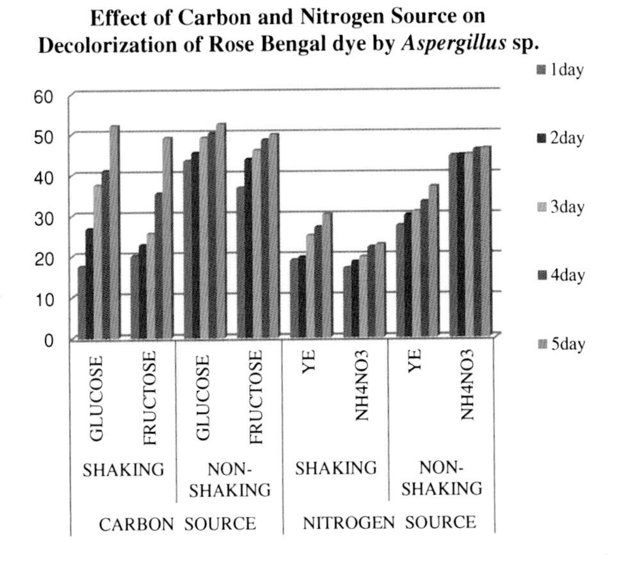

Effect of pH on the Rate of Decolorization of Congo Red and Rose Bengal: The effect of pH at 4, 5, 6, 8, 9 on the rate of decolorization of Congo red by *Aspergillus* sp. showed that under static conditions the maximum decolorization was 84 % at pH 6 and at pH 4 and 5 the decolorization was 71 and 68 % and under shaking conditions the maximum decolorization was only 40 % whereas pH 8 and 9 showed no decolorization and the rate of decolorization of Rose bengal by *Aspergillus* sp. showed that under static conditions with the increase in pH the dye decolorization has increased till pH 8 and decreased, In pH 8 the maximum decolorization was 40 % and under shaking conditions the maximum decolorization was 51 % in pH 4 where as pH 9 showed no decolorization.

Effect of Temperature on the Rate of Decolorization of Congo Red and Rose Bengal: The effect of temperatures 25, 30 and 40 °C on the removal of Congo red by *Aspergillus* sp. showed that in static conditions as the temperature increased Congo red dye decolorization also increased. At temperature 40 °C the decolorization was 87 % whereas at temperatures 25 and 30 °C the decolorization of dye was 80 and 84 % respectively and under shaking conditions 90 % decolorization was observed at temperature 30 °C and 86 % decolorization at temperatures 25, 40 °C and the effect of temperatures on the removal of Rose bengal by *Aspergillus* sp. at 25 °C was only 41 % whereas at temperatures 30 and 40 °C the decolorization of dye was 38 and 34 % respectively and under shaking conditions as the temperature increased Rose Bengal dye decolorization also increased. At temperature 40 °C the decolorization was only 25 %.

Effect of P^H on Decolorization of Congo red by *Aspergillus Sp.*

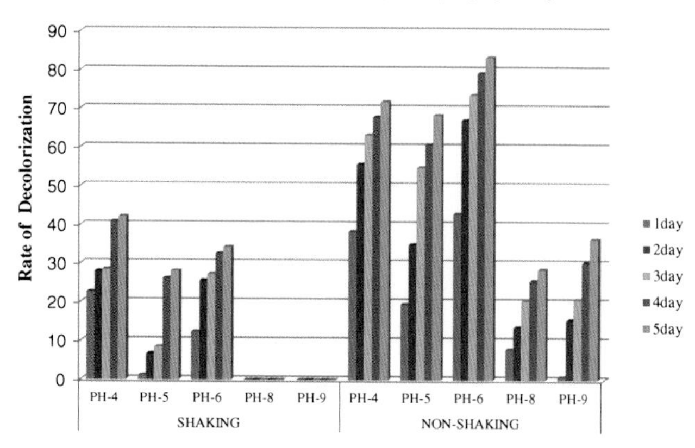

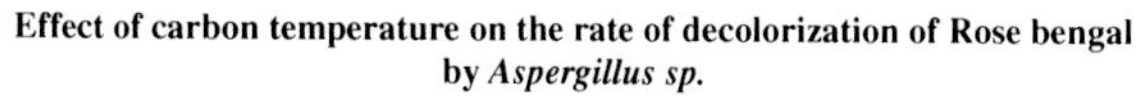

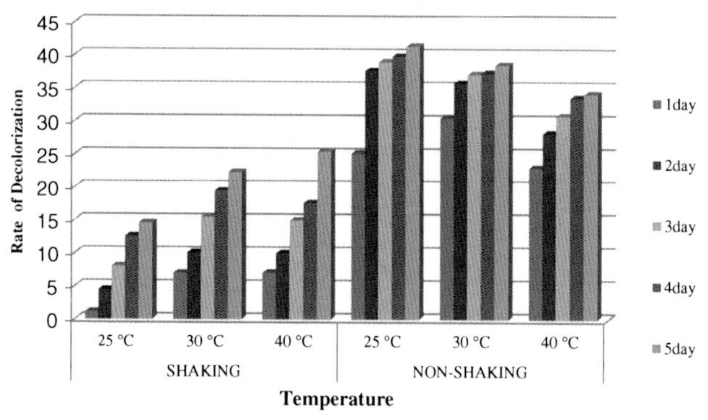

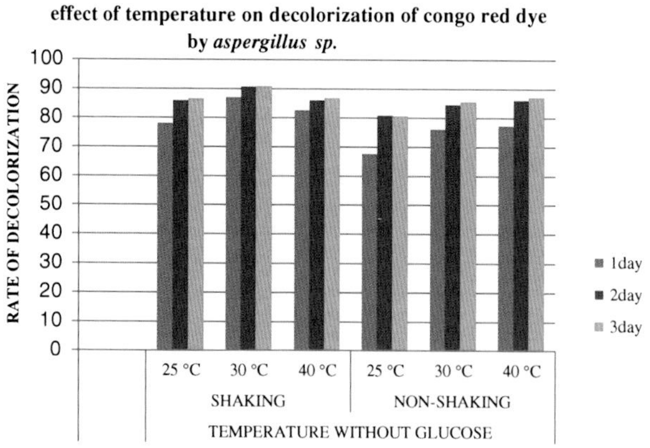

effect of temperature on decolorization of congo red dye by *aspergillus sp.*

RATE OF DECOLORIZATION

100
90
80
70
60
50
40
30
20
10
0

25 °C 30 °C 40 °C 25 °C 30 °C 40 °C

SHAKING NON-SHAKING

TEMPERATURE WITHOUT GLUCOSE

1day
2day
3day

3.5 Biosorption Experiment

In the First experiment with different (1, 3 and 5 mg) concentrations of Congo red and Rose Bengal dye and single (500 mg) amounts of inactive dead mycelia mat of *Aspergillus* sp. Indicated that binding of the dye to dead mycelia mat occurred rapidly with more than 50 % within 1 day and the dye absorbed to the mat increased with increase in the dye concentration. A maximum of 93 and 90 % decolorization was attained at 5 mg Congo red and Rose Bengal concentrations and in the more than 50 % within 1 day and the dye absorbed to the mat increased with increase in the dye concentration. A maximum of 93 and 90 % decolorization was attained at 5 mg Congo red and Rose Bengal concentrations and in the second experiment with single (3 mg) concentrations of Congo red and Rose Bengal dye and different (350, 550 and 750 mg) amounts of inactive dead mycelia mat of *Aspergillus* sp. Indicated that binding of the dye to dead mycelia mat occurred rapidly with more than 50 % within 1 day and the dye absorbed to the mat increased with increase in the mycelia mat. A maximum of 92 and 88 % decolorization was attained at 750 mg of dead mycelia mat the mycelia mat concentration.

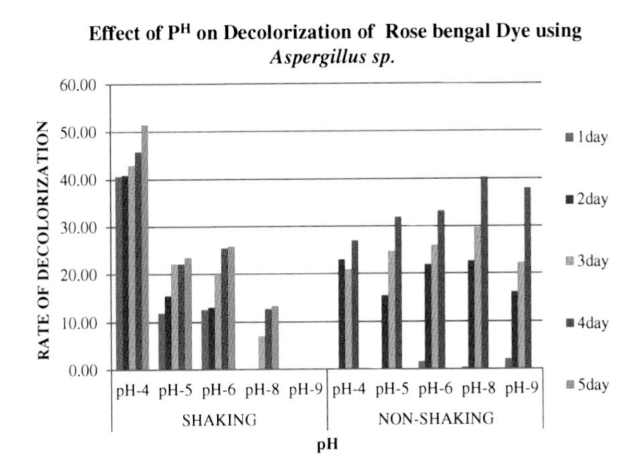

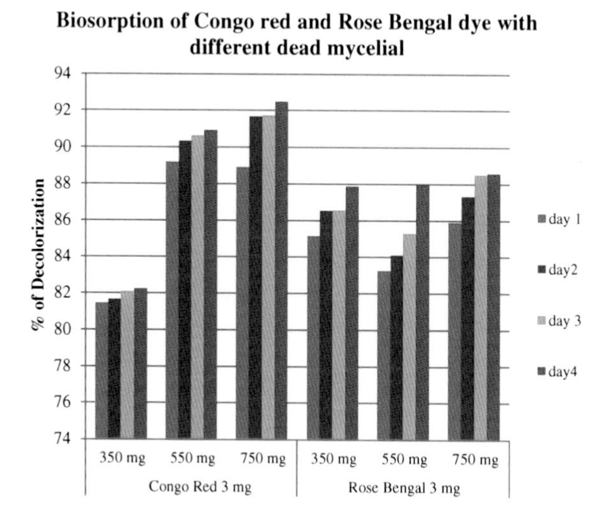

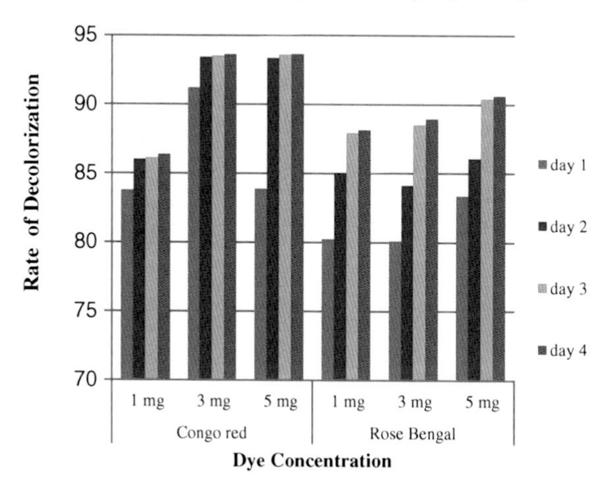

3.6 Conclusion

In the present study percent decolorization achieved by metabolically inactive mycelium was equal to that attained by the live mycelium. The rapidity of decolorization by autoclaved mycelium indicates a phenomenon of biosorption.

Autoclaving might have increased the biosorption capacity of fungal biomass [8]. Biosorption of the reactive dye into interior region of spherical biomass pellets of *Aspergillus* sp. was in agreement with the observation made in the present study [9]. Binding of the dye to dead mycelia mat occurred rapidly with more than 50 % within 1 day and decolorization achieved by metabolically inactive mycelium was equal to that attained by the live mycelium. Decolorization by live cells involves more complex mechanisms such as intracellular, extracellular oxidases and biosorption, than the dead cells. The process involving living cells is closely related to the operational conditions, such as nutritional requirements, the influent concentration and toxicity. In contrast decolorization involving dead biomass is easier to operate and dead cells may possess higher biosorption capacity in certain conditions. Thus fungal decolorization is a promising alternative to replace present treatment processes. However, using fungal biomass to remove color in dye waste water is still in research stage. More studies are needed to develop a practical application.

References

1. Kumar MNV, Sridhari TR, Bhavani KD, Dutta PK (2001) Trends in color removal from textile mill effluents. Biotechnology 56:81–87
2. Bhattacharyya KG, Sharma A (1998) Adsorption characteristics of the dye, Brilliant Green. Bioresour Technol 77:25–34
3. Husseiny M (2008) Biodegradation of the reactive and direct dyes using Egyptian isolates. J Appl Sci Res 4(6):599–606
4. Vandevivere PC, Bianchi R, Verstraete W (1998) Treatment and reuse of wastewater from the textile wet-processing industry: review of emerging technologies. J Chem Technol Biotechnol 72:289–302
5. Robinson T, Chandran B, Nigam P (2001) Studies on the production of enzymes by white-rot fungi for the decolorisation of textile dyes. Enzym Microb Technol 29:575–579
6. Slokar YM, Majcen Le Marechal A (1998) Methods of decoloration of textile waste waters. Dyes Pigm 37:335–356. ISSN 0143-7208
7. Ramya TNC, Karmodiya K, Surolia A, Surolia N (2007) 15-Deoxyspergualin primarily targets the trafficking of apicoplast proteins in *Plasmodium falciparum*. J Biol Chem 282:6388–6397
8. Fu Y, Viraraghavan T (2001) Fungal decolorization of dye wastewaters: a review. Bioresour Technol 79:251–262
9. Aretxaga A, Romero S, Sarra M, Vicent T (2001) Adsorption step in the biological degradation of a textile dye. Biotechnol Prog 17:664–668

Chapter 4
Anti-cancer Activity of Selected Seaweeds Against HeLa, K-562 and MDA-MB Cell Lines

Ilahi Shaik, A. Shameem and P. Sasi Bhushana Rao

Abstract Anticancer activity of marine macro algae (seaweeds) *Sargassum wightii, Ulva fasciata* and *Gracillaria corticata* against HeLa, K-562 and MDA-MB cell lines was studied. During the present investigation crude extracts of seaweeds were prepared using soxhlet apparatus. Hexane, Butanol and Methanol were used as solvents. Crude extracts were dissolved in DMSO. In vitro anticancer activity of seaweeds at various concentrations (12.5–200 µg/ml) was studied against the chosen cell lines using MTT assay (3-(4,5-dimethylthiazol-2-yl)-2, 5-diphenyltetrazolium bromide, a yellow tetrazole). Butanolic extract of *Gracillaria corticata* (200 µg/ml) has shown greater anticancer activity than hexane and methanolic extracts of *G. corticata* and butanol, hexane and methanolic extracts of *Sargassum wightii* and *Ulva fasciata*.

Keywords Seaweeds · HeLa · K-562 and MDA-MB cell

4.1 Introduction

Cancer is one of the most dreaded diseases accounting for death of more than 3,500 people per million population around the world. Although Chemotherapy is the best option available for the treatment of cancer, its severe side effects cannot

I. Shaik (✉) · P. Sasi Bhushana Rao
Department of Biochemistry, Gayatri Vidya Parishad Degree College (Autonomous), M.V.P. Colony, Visakhapatnam 530 017, Andhra Pradesh, India
e-mail: ilahi.mohammad@gmail.com

P. Sasi Bhushana Rao
e-mail: psbrao@gmail.com

A. Shameem
Department of Marine Living Resources, Andhra University, Visakhapatnam 530 003, Andhra Pradesh, India

A. Kumar, *Biotechnology and Bioforensics*, Forensic and Medical Bioinformatics,
DOI: 10.1007/978-981-287-050-6_4,

be ignored. This has necessitated the search for alternative anticancer drugs having better efficacy with minimum side effects. This search turned the focus towards the marine macro algae which are tremendous reserve sources of pharmacologically efficient compounds [1, 2].

Marine macro algae, the seaweeds, constitute significant proportion of plant biomass of the marine ecosystem and reside at the intertidal zone, which is highly dynamic in nature. Till date, more than 2,400 marine bioactive compounds have been isolated from the seaweeds of subtropical and tropical populations [3]. Recent studies reveal that they possess antiviral, antibacterial, antifungal and antitumor activity [4]. Depending on the pigments seaweeds are broadly classified into chlorophyta, pheophyta and rhodophyta. The seaweeds *Sargassum wightii*, *Ulva fasciata* and *Gracillaria corticata*, belonging to the family pheophyceae, chlorophyceae and rhodophyceae respectively were tested for their anticancer activity. Crude extracts of the above three seaweeds prepared in hexane, butanol and methanol were used against HeLa-2 (cervical Cancer), K-562 (leukemia) and MDA-MB-231 (Breast cancer) cell lines using MTT assay during the investigation.

4.2 Materials and Methods

4.2.1 Sample Collection and Preparation

Rao and Sreeramulu [5] described the seaweeds of Visakhapatnam coast. Selected sea weeds were procured from the intertidal region of Bay of Bengal at Visakhapatnam, and the samples were identified by following the standard procedure [6, 7]. They were brought to the laboratory in polythene bags and thoroughly cleaned with sea water followed by distilled water to remove sand, salt and epiphytes. Samples were shade dried and coarse powdered.

4.2.2 Preparation of Extracts

Sample and the solvents were taken in 1:10 ratio and soxhaltion had been performed by using hexane, butanol and methanol. Crude extracts were concentrated by rotary evaporator at 45–50 °C. The resulting extracts were dissolved in Di Methyl Sulphoxide (DMSO) and kept in freezer for further use.

4.2.3 Cell Lines and Culture Condition

HeLa, K-562 and MDA-MB-231 cell lines were purchased from National Council for Cell Sciences (NCCS) Pune. Cell lines were re-suspended in Minimum

Essential Medium (MEM) medium and culture flasks were incubated in 5 % CO_2 incubator at 37 °C for 2–3 days.

4.2.4 MTT (3-(4, 5-Dimethylthiazol-2-yl)-2, 5-Diphenyltetrazolium Bromide) Assay

Above three cell lines were added into 96 wells plate at a concentration of 2×10^4 cells/ml and to each well 20 μl of MEM was added. The plates were incubated over night in 5 % CO_2 incubator at 37 °C. Algal crude extracts of different concentrations of 12.5, 25, 50, 100, 200 μg/ml were prepared in DMSO. The cell lines were treated with 50 μl of the algal crude extracts of above different concentrations. Blank well had only media. Control cell lines were run without addition of crude extracts. All the plates were incubated for 24 h. After incubation 20 μl of freshly prepared MTT solution (5 mg/ml in Phosphate Buffer Saline) was added in each well and incubated for 4 h to allow the MTT to be metabolized. Media was discarded after incubation. Formozan (metabolic product of MTT) that was left in the well was re-suspended in 200 μl DMSO. The plates were placed on a shaking table at 150 rpm for 5 min for thorough mixing of Formozan into DMSO. The amount of MTT-Formozan that is directly proportional to the number of living cells was determined by reading the optical density (OD) at 420 nm.

4.2.5 Calculation

Percentage of cell inhibition was calculated by using the formula given below:

Percentage of cell survival (CS %) = AT − AB/AC − ABX100
Percentage of cell inhibition (CI %) = 100 − CS %
AT = absorption of test
AB = absorption of blank
AC = absorption of control.

4.3 Results

Results of in vitro anticancer activity of hexane, butanol and methanolic extracts of *S. wightii*, *U. fasciata* and *G. corticata* against HeLa, K-562 and MDA-MB Cell Lines, are given in Tables 4.1, 4.2 and 4.3. The same are shown in Figs. 4.1, 4.2 and 4.3. Crude extracts showed increase in inhibition of cell lines (antitumor activity) with increase of concentration. At minimal (12.5 μg/ml) concentration

Table 4.1 Percentage of inhibition against HeLa cell lines by different concentrations of crude extracts of the three seaweeds

Conc. in μg/ml	*G. corticata* percentage of inhibition			*S. wightii* percentage of inhibition			*U. fasciata* percentage of inhibition		
	Butanol	Hexane	Methanol	Butanol	Hexane	Methanol	Butanol	Hexane	Methanol
12.5	10.2	5.73	5.44	2.26	1.17	3.71	7.08	5.79	3.5
25	15.54	22.92	13.5	10.17	11.17	5.3	13.72	6.94	4.79
50	21.36	24.48	15.7	12.43	16.96	10.06	16	10.99	5.39
100	26.7	25	19.5	18.08	17.55	28.58	19.43	21.97	15.56
200	40.78	35.94	33.1	25.99	25.15	29.63	36	24.28	27.54

Table 4.2 Percentage of inhibition against MDA-MB 231 cell lines by different concentrations of crude extracts of the three seaweeds

Conc. in μg/ml	*G. corticata* percentage of inhibition			*S. wightii* percentage of inhibition			*U. fasciata* percentage of inhibition		
	Butanol	Hexane	Methanol	Butanol	Hexane	Methanol	Butanol	Hexane	Methanol
12.5	11	1.6	8.61	5.16	2.95	0.86	12.7	19.53	17.82
25	13.13	5.18	16.49	15.47	11.35	1.72	13.1	25	25.46
50	21.64	10.36	19	19.59	12.61	3.44	17.46	31.85	26.55
100	23.54	17.53	32.26	25.09	30.68	25.76	37.31	40.07	42.91
200	56.03	51.4	52.69	41.93	32.36	41.64	50.8	50.69	54.91

Table 4.3 Percentage of inhibition against K-562 cell lines by different concentrations of crude extracts of the three seaweeds

Conc. in μg/ml	*G. corticata* percentage of inhibition			*S. wightii* percentage of inhibition			*U. fasciata* percentage of inhibition		
	Butanol	Hexane	Methanol	Butanol	Hexane	Methanol	Butanol	Hexane	Methanol
12.5	33.86	22.72	9.46	25.07	14.84	11.08	1.33	28.09	26.86
25	36.49	27.98	8.79	31.54	29.36	18.04	18.61	31.18	35.72
50	41.47	34.91	30.75	33.43	30.65	25.64	27.25	33.99	37.43
100	45.67	43.22	31.09	37.47	27.75	34.18	43.19	40.17	41.72
200	57.22	50.14	39.19	50.95	42.59	45.89	44.52	49.44	49.71

cell inhibition was observed to be below 10 %. At 50–200 μg/ml a 20–50 % greater inhibition was noticed. Maximum inhibition was observed at 200 μg/ml.

Butanolic, hexane and methanolic extracts of *G. corticata* and *U. fasciata* had shown more than 50 % inhibition against MDA-MB 531. Butanol and hexane extracts of *G. corticata* have inhibited 50 % of K-562 and 40 % of HeLa cell line growth respectively. Effect of *S. wightii* and *U. fasciata* is below 30 %.

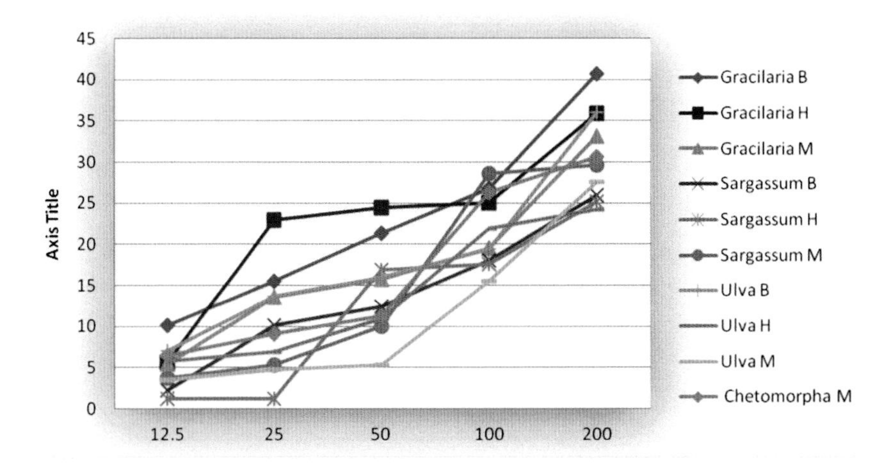

Fig. 4.1 Anticancer activity on HeLa cell line

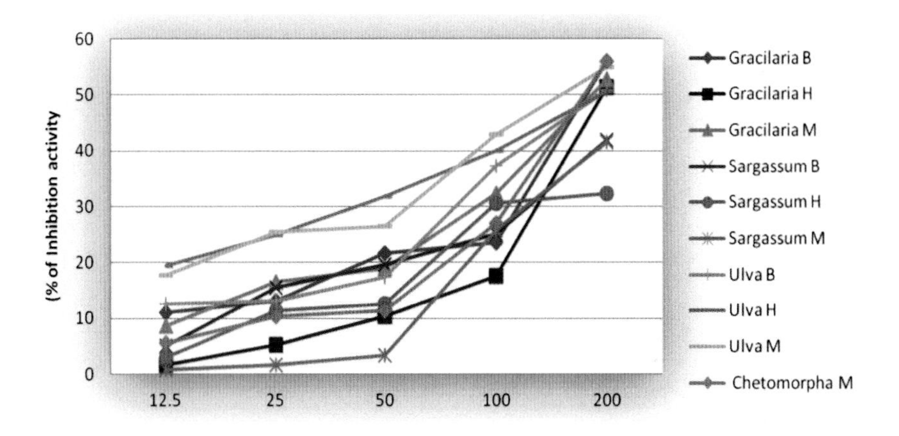

Fig. 4.2 Anticancer activity on MDA-MB-231 cell line

4.4 Discussion

Marine environment is a source of potential bioactive compound [8–12]. The green algae are known for their antibacterial, antiviral, antioxidative, and anti-proliferative properties [13–16]. In our observation also *U. fasciata* has shown 50 % cell inhibition against K-562 and MDA-MB 531 cell lines where as the

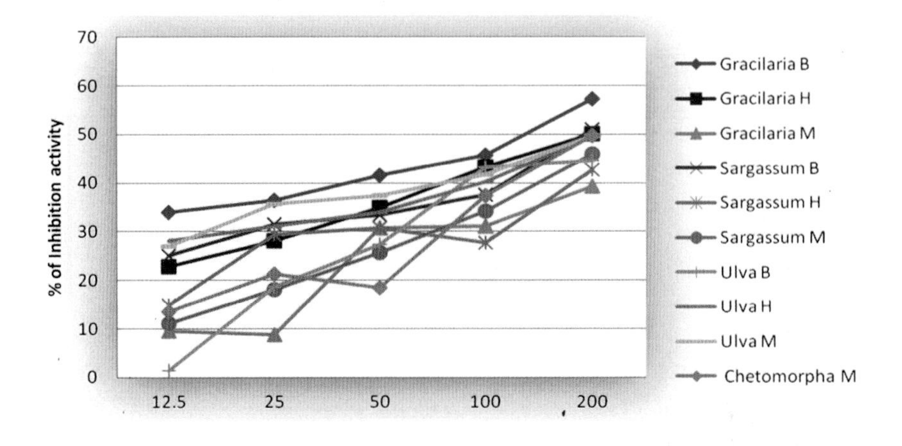

Fig. 4.3 Anticancer activity on K562 cell line

inhibitory effect against HeLa cell lines is very low. Our data reveals that butanolic and hexane extracts of *U. fasciata* are found to be more effective against K-562 while methanolic extract is effective against the MDA-MB 531 cell lines.

Brown algae which possesses sulphated polysaccharides [17, 18] and fucosidian are very good source of anticoagulant, antibacterial and antioxidant as well as antitumourogenic compounds. Different organic and aqueous extracts of *Sargassum* sp. were reported for their anticancer properties [19]. Present data showed that Butanolic and methanolic extracts of *S. wightii* are effective against K-562 and MDA-MB 531 cell lines. Hexane extract has shown the minimum cytotoxicity. The compounds which are responsible for biological activity might be moderate polar in nature.

Earlier reports show that *Gracillaria* sp. ethanolic, chloroform and aqueous extracts inhibited cell viability [20]. In our study *G. corticata* organic solvent extracts also inhibited more than 50 % of the cell proliferation of HeLa, K-562, and MDA-MB 531 cell lines.

4.5 Conclusion

Reactive oxygen species (ROS) may be one of the causative agents of cancer that damages DNA leading to mutagenesis. Antioxidants are the compounds which scavenge ROS [21]. Previous literature suggests the presence of poly phenols [22], carotenoids, sulfated polysaccharides, terpenoids, flavonoids, phycocolloids, alkaloids, sterols [23] in marine algae which act as the potent source of antioxidants. In our studies *Gracillaria corticata* has been the most effective inhibitor of cancer cell lines. Therefore *Gracillaria corticata* may be containing the above substances which are nonpolar and moderate polar. Further investigation has to be done to identify, isolate, and evaluate the specific compounds which are responsible for anticancer activity and in vivo studies have to be carried out to check their cytotoxicity on normal cells.

Acknowledgment The first author thanks the Management and the Principal of G.V.P. College for Degree and P.G. Courses, for the encouragement and support, Prof. A. Shameem and Dr. P. Janaki Ram, Department of Marine Living Resources, Andhra University for their guidance and Prof. P. Sasi Bhushana Rao for guidance and critically going through the manuscript and suggestions.

References

1. Blunden G (1993) Marine algae as sources of biologically active compounds. Interdisc Sci Rev 18:73–80
2. Smit AJ (2004) Medicinal and pharmaceutical uses of seaweed natural products: a review. J Appl Phycol 16:245–262
3. Manilal A, Sujith S, Kiran GS, Selvin J, Shikar C (2009) Cytotoxic potentials of red alga, *Laurencia brandenii* collected from the Indian Coast. Global J Pharmacol 3:90–94
4. Harada H, Noro T, Kamei Y (1997) Selective antitumor activity in vitro from marine algae from Japan coast. Biol Pharm Bull 20:541–546
5. Uma Maheswara Rao M, Sreeramulu T (1970) An annotated list of the marine algae of Visakhapatnam (India). Bot J Linn Soc 63:23–45
6. Dharallear VK, Kavlekar DP (2004) Seaweeds: a field manual. National Institute of Oceanography, Goa
7. Sambarmurty AVSS (2005) A text book of algae. I.K. International Pvt Ltd, New Delhi, pp 65–97
8. Hashimoto Y (1979) Marine toxins and other bioactive marine metabolites. Japan Scientific Societies Press, Tokyo, p 369
9. Naqvi SWA, Kamat SY, Fernandes L, Reddy CVG, Bhakuni DS, Dhawan BN (1980) Screening of some marine plants from the Indian coast for biological activity. Bot Mar 24(1):51-55
10. Barbier M (1981) Marine chemical ecology: the roles of chemical communication and chemical pollution. In: Scheuer PJ (ed) Marine natural products: chemical and biological prospectives, vol 4. Academic Press Inc., New York, pp 148–186
11. Scheuer Paul J (ed) (1982) Marine natural products: chemical and biological prospectives, vol 1–5. Academic Press Inc., New York
12. Faukner DJ (1984) Marine natural products: metabolites of marine algae and herbivorous marine mollusks. Nat Product Rep 1:251
13. Hatano T, Kagawa H, Yasuhara T (1988) Evaluation of marine algae *Ulva lactuca*. L as a source of Natural Preservative Ingredient. American-Eurasian J Agric Environ Sci 3:434–444
14. Mayer AMS, Lehmann VKB (2000) Marine pharmacology in 1998: Marine compounds with antibacterial, anticoagulant, anti-inflammatory, anthelmintic, antiplatelet, antiprotozoan, and antiviral activities; with actions on the cardiovascular, endocrine, immune, and nervous systems; and other miscellaneous mechanisms of action. Pharmacologist 42:62–69
15. Awad NE (2000) Biologically active steroid from the green alga *Ulva lactuca*. Phytother Res 14:641–643
16. Chakraborty K, Paulraj R (2010) Sesquiterpenoids with free-radical-scavenging properties from marine macroalga Ulva fasciata Delile. Food Chem 122:31–41

17. Ribeiro-do-Valle RM (2005) Antiangiogenic and antitumoral properties of polysaccharide isolated from the seaweed *Sargassum stenophyllum*. Cancer Chemother Pharmacol 56:436–446
18. Albano RM, Pavao MS, Mourao PA, Mulloy B (1990) Structural studies of a sulfated L-galactan from *Styela plicata* (Tunicate): analysis of the smith-degraded polysaccharide. Carbohydr Res 208:163–167
19. Zandi K, Ahmadzadeh S, Tajbakhsh S, Rastian Z, Yousefi F, Farshadpour F, Sartavi K (2007) Anticancer activity of *Sargassum oligocystum* water extract against human cancer cell lines. Phytother Res 21:170–175
20. Sundaram M, Patra S, Maniarasu G (2012) Antitumor activity of ethanol extract of *Gracilaria edulis* (Gmelin) Silva on Ehrlich ascites carcinoma-bearing mice. J Chin Integr Med/Zhong Xi Yi Jie He Xue Bao 10(4):430–435
21. John MC (1995) Lipid peroxidation and antioxidants as biomarkers for tissue damage European Beckmen Conference. Clinical Chem 41(12):1819–1828
22. Han X, Shen T, Lou H (2007) Dietary polyphenols and their biological significance. Int J Mol Sci 8:950–988
23. Abd El-Baky HH, El Baz FK, Gamal SE, Baroty I (2008) Evaluation of marine alga *Ulva lactuca* L. as a source of natural preservative ingredient. American-Eurasian J Agric Environ Sci 3(3):434–444

Chapter 5
Predisposition Factors of Type II Diabetes Mellitus and Related Complications

Alice Jayapradha Cheekurthy, C. Ram Babu, Amit Kumar and K. Surendrababu

Abstract Diabetes mellitus is considered to be one of the leading causes of death in adult and children. The most challenging complex heterogeneous group of conditions is prevalent in developing countries as well as the western world is T II DM. Rise in upcoming cases of diabetes has posed it as a global health problem. The predisposition factors are responsible for identifying at risk individuals for this polygenic disorder. Several controllable and non-controllable predisposition factors are responsible for identification of at-risk individuals to T II DM. The complex interactions between genetic and environmental and biochemical predisposition factors unmask the advancement of the disease. This review gives brief description and table form of the work done by the researchers for predisposition of Type II diabetes and some of the related complications among Asian Indian phenotypes and world wide.

Keywords Type II diabetes mellitus (T II DM) · Predisposition factors · Asian Indian phenotypes

A. J. Cheekurthy (✉)
Department of Biochemistry, Acharya Nagarjuna University, Guntur 522 510, Andhra Pradesh, India
e-mail: a.jayapradha@dnares.in

C. Ram Babu
Acharya Nagarjuna University, Guntur 522 510, Andhra Pradesh, India

A. Kumar · K. Surendrababu
BioAxis DNA Research Centre (P) Limited, Hyderabad 500 068, Andhra Pradesh, India

A. Kumar, *Biotechnology and Bioforensics*, Forensic and Medical Bioinformatics,
DOI: 10.1007/978-981-287-050-6_5,

5.1 Introduction

Diabetes mellitus is one of the first diseases described. The term "diabetes" or "to pass through" was first used in 230 BCE by the Greek physician Apllonius of Memphis. The term "mellitus" which means "from honey" was added by "Thomas Willis" in the late 1600s to separate it from diabetes insipidus. All the cells in our body need glucose to work normally. Disorder that disrupts the way your body uses glucose i.e. metabolic defect in the body's ability to convert glucose to energy. Based on the etiology the three main kinds of diabetes mellitus are Type I (juvenile), Type II (maturity onset), and Gestational diabetes. The most common type of diabetes is T II DM also called adult onset or non-insulin dependent characterized by insulin resistance. Distinction between Type I diabetes and Type II diabetes was first clearly done in 1936 [1].

Global burden of Diabetes [2] (Courtesy IDF Diabetes Atlas)

- **366 million** people have diabetes in 2011; by 2030 this will have risen to **552 million**
- The number of people with type 2 **diabetes is increasing** in every country
- **80 %** of people with diabetes live in **low- and middle-income countries**
- The **greatest number** of people with diabetes are between **40 and 59** years of age
- **183 million** people (50 %) with diabetes are **undiagnosed**
- Diabetes caused **4.6 million deaths** in 2011
- Diabetes caused at least **USD 465 billion dollars** in healthcare expenditures in 2011; **11 % of total healthcare expenditures** in adults (20–79 years)
- **78,000 children** develop **type 1 diabetes** every year

Diabetes mellitus is the most prevalent non-communicable metabolic disease affecting more than 100 million people not only in urban populations of India but also in rural populations. India is the diabetic capital of World owing to having large number of diabetics than any other part of the world. The number of people with diabetes in India currently around 40 million is expected to rise to around 70 million by 2025 [3]. Among the three countries with maximum number of diabetics India tops the list with 79.4 million in 2030; where as the projections for China and the US are 42.3 million in 2030 and 30.3 million in 2030 respectively. If preventive measures are not taken according to International Diabetes Federation, the prevalence of diabetes for all age-groups worldwide will be increased to 4.4 % in the year 2030. The total number of people affected with diabetes will rise to 366 million in 2030 [4]. The future projections of diabetes mellitus prevalence among the other countries (The IDF Diabetes Atlas) for the year 2030 are depicted below.

Countries with highest number of estimated cases of T II DM for the year 2030 [6]

Ranking	Country	T II DM people
1	India	79.4
2	China	42.3
3	U.S.	30.3
4	Indonesia	21.3
5	Japan	13.9
6	Pakistan	11.3
7	Russia	11.1
8	Brazil	8.9
9	Italy	7.8
10	Bangladesh	6.7

This review describes global studies on of diabetes over years. Earliest study on prevalence of diabetes in India done by documentation record is in Kolkata in 1938 [5] followed by studies in 1959 in Mumbai [6]. More than 70 % of the population resides in rural area where studies revealed much higher prevalence that may be expected which started in early 70s. First ever Studies in rural areas of different parts of the country based on the WHO criteria were done on farmers following traditional life style and are dependent on agricultural economy [7]. The prevalence was higher among women above 50 years of age and that of men was irrespective of age. Prevalence of known diabetes was 6.1 % in all subjects aged 40 or over and rose to 13.3 % in the age group 50–59 years [8] as reported in door to door survey in 80s. Glucose intolerance is present 11.5 % of the rural South Indian population aged 40 years or over is the finding of 1994 study [9]. Very high levels of diabetes in comparison to rural population have been reported in urban areas of India [10] and other parts of the world in the age group 45–64 years and 65 years respectively. T II DM is designated as disease of elderly and middle aged but in recent years shift in the age of onset of diabetes is observed as major issue of worry in children as well as adolescents. The no of reports of children affecting from diabetes is increasing year by year day. Before 1990, there were only 2 reports; between 1990 and 1994, 4 reports; 1995–1999, 12 reports; and between 2000 and 2003, 53 reports and the number goes on increasing. A few of the findings are in the table below.

Children and adolescents affected with diabetes mellitus

S. No.	Population	Area	Finding	References
1.	Adolescents and children	North America	Sixfold increase	[11]
2.	Adolescents	Greater Cincinnati	Obesity and strong family history	[12]
3.	Pima Indians	Arizona	Increased from 0.3 to 1.2/100,000/year before 1992 to 2.4/100,000/year in 1994	[13]

5.1.1 *Predisposition Factors*

Several controllable and non-controllable predisposition factors are responsible for identification of at-risk individuals to T II DM. These broadly classified into Environmental Predisposition, Biochemical Predisposition factors and Genetic Predisposition factors.

Environmental Predisposition Factors Asian Indian people are more prone to diabetes than European descent people [10, 14]. Major environmental factors that contribute long-term complications associated with the of diabetes [15]. The highest rates of T II DM are found among Native Americans, particularly the Pima Indians who reside in Arizona in the US, and in natives of the South Pacific islands, such as Nauru [4]. There is high prevalence of T II DM among Europeans, Americans, Chinese, and Asian Indians. The variation is due to increased insulin resistance according to inter ethnic difference. A factor such as obesity, sedentary lifestyle, diet rich in animal products, and aging has attributed to tremendous increase in T II D Obesity is an established risk factor for T II DM [16].

Comparisons of characteristics of the three main race and ethnic groups

S. No.	Population	Percentage affected (%)	Reference
1.	Hispanic Caucasians	74.4	[17]
2.	Non-hispanic African–African–Americans	15.0	[17]
3.	Mexican–Americans	5.8	[17]

5.1.2 *Biochemical Predisposition Factors*

Endothelial dysfunction (ED) is associated with the presence of atherosclerosis [18]. There is association of many novel biochemical Predisposition factors with incidence of diabetes. The poor glycaemic control, a longer duration of diabetes results in the complications of the Type II diabetes [19].

Genetic Predisposition Factors Genetic predisposition to the disease is indicated by its characteristic feature to run in a family. Progress in gene identification for genomic DNA sequence variation more common responsible, multifactorial forms of T II DM has been slower may genes are identified and a large number of researchers have given lot of contribution for the study of Genetic Predisposition of T II DM. This review tries attempts to summarize work done by various researchers.

Gene	Year	Population affected	References
PPARG	2005	Diabetic retinopathy in the Slovene population (Caucasians)	[20]
	2000	Increases risk	[21]
PTPNI	2006	*Association with obesity and T II DM in French population	[22]
	2004	*Association with T II DM in two independently ascertained collections of Caucasian subjects	[23]
ENPP1/PC-1	2011	Meta-analysis shows association with obesity in European adult populations	[24]
GCK	2013	Associations vary in different ethnic populations	[25]
KCNJ11(Kir6.2)	2007	*Variant *KCNJ11* E23K in the five ethnic groups: Caucasians, African Americans, Hispanic Americans, Asian Americans, and American Indians Associated	[26]
			[27]
SUR1 (ABCC8)	2007	*T II DM and blood pressure levels in the Japanese population *Associated with T II DM	[28]
CAPN10	2000	*Association with type 2 diabetes in Mexican Americans and a Northern European population from the Botnia region of Finland	[29]
TCF7L2	2009	*Role in β cell function	[30]

Association studies of candidate gene variants have revealed the affect of gene polymorphism. A genotype of the Gly482Ser polymorphism in the PPARGC1 gene might be a risk factor for diabetic retinopathy Polymorphism of the SUR1 gene and the E23K polymorphism of the Kir6.2 gene predicted the conversion from IGT to T II DM Genes Predict the Conversion from Impaired Glucose Tolerance to Type 2 Diabetes. Lysine variant in *KCNJ11* E23K leads to diminished insulin secretion in individuals with IGT. Calpain-10 polymorphisms impair the function of pancreatic beta-cells in humans ENPP1 Q121 variant modestly increases the risk of T II DM diabetes and is associated with obesity in adult Europeans. Mild fasting hyperglycaemia caused by heterozygous *GCK* mutations. Late onset autosomal dominant diabetes mellitus is due a role for HNF1A PTPN1 gene, inactivates the insulin signal transduction cascade by dephosphorylating phosphotyrosine residues. Common variants in the gene that encodes the transcription factor 7-like 2 (TCF7L2) have been strongly associated with type 2 diabetes.

5.1.3 Complications

Today T II DM is one of the most significant public health problems faced globally, is the sixth leading cause of death. T II DM duration along with genetic predisposition factors has been associated with complications [19]. They can affect every part of the body manifesting in different ways for different people. People with diabetes have twice the risk of death. The reason behind these deaths is not diabetes itself but the complications arising from this are responsible. Diabetes is a complex disease that risks individual for other complications. They can affect every part of the body manifesting in different ways for different people. Many millions are exposed to the increased risk of Complications like Cardiovascular disease resulting in dyslipidemia, hypertension. Predisposition factors for sudden cardiac death that may be less prevalent in Hispanics than non-Hispanic Whites include prevalent myocardial infarction in Mexican Americans, along with heavy cigarette smoking. In two studies African Americans had higher cardiac arrest rates and poorer outcomes following cardiac arrest than Whites. Single gene mutations that result in hypertension are angiotensinogen gene human G protein B3 subunit gene. Musculoskeletal disorders are associated with people affected with T II DM Neuropathy in which there is dysfunction of the vascular endothelium. The skin is a potentially invaluable tool for understanding certain diabetic complications. The development of several skin manifestations in insulin-dependent patients seems to be related to duration of diabetes and to development of diabetic microvascular complications. *Obesity* The higher serum concentrations of TNFα, nitric oxide (NO) and IL10 concentration is associated with overweight and obesity in young adolescents. *Alzheimer's disease* APOE is involves redistribution and in the mobilization of cholesterol in the periphery and during neuronal growth and repair. Many more are encountered commonly among those suffering from T II DM.

A few complications related to T II DM

S. No.	Complication	Affecting organ
1.	Cardiovascular disease	Heart
2.	Neuropathy	Skin
3.	Obesity	Adiposity of skin
4.	Musculoskeletal disorder	Muscles and skeletal system

5.2 Conclusion

The Review summarizes a few of the available papers on Predisposition factors for Type II Diabetes Mellitus and complications related to it. Diabetes is a disease which develops slowly so the number of persons affected is far more than actually diagnosed. Population studies have revealed the presence of diabetes in the people that have not been previously diagnosed. The complications related to Type II diabetes mellitus are due to the prolong duration Scientific evidence shows is a life style disorder no cure for the T II DM. By making a small modification in your life style (weight reduction, more exercise) has been shown to forestall, and even prevent, development of diabetes in susceptible individuals as it will help to prolong the T II DM onset and you can live a longer and healthier life preventing serious problems. Discovering which genetic variations is of both theoretical and practical importance and help to physician because it could help for the better treatment. The proven susceptibility genes will be themselves targets for pharmaceuticals for the treatment. Predisposition is definitely a help to identify the disease occurrence in different populations. More beneficial results could be gained by combining information obtained from several different predisposition factors and not relying on the single.

References

1. Himsworth HP (1936) Diabetes mellitus: its differentiation into insulin-sensitive and insulin-insensitive types. Lancet 227(5864):127–130
2. International Diabetes Federation (2011) IDF Diabetes Atlas, 5th edn. Brussels, International Diabetes Federation, Belgium
3. Sicree R, Shaw J, Zimmet P (2006) Diabetes and impaired glucose tolerance. In: Gan D (ed) Diabetes atlas, 3rd edn. International Diabetes Federation, Belgium, pp 15–103
4. Wild S et al (2004) Global prevalence of diabetes estimates for the year 2000 and projections for 2030. Diabetes Care 27(5):1047–1053
5. Chakravarthy A (1938) A study of diabetes and its treatment with special reference to Bengalis and their diet. Indian Med Rec 58:65
6. Patel JC, Dhirawani MK (1959) Incidence of diabetes in Bombay: analysis of 8 years' case records of a general hospital. Indian J Med Sci 12:10
7. Jaya Rao KS et al (1972) A survey of diabetes mellitus in a rural population of India. Diabetes 21:1192–1196
8. Rao PV et al (1989) The Eluru survey: prevalence of known diabetes in a rural Indian population. Diabetes Res Clin Pract 7(1):29–31
9. Blots ML et al (1994) Impaired glucose tolerance and diabetes mellitus in a rural population in South India. Diabetes Clin Pract 24(1):47–53
10. Ramachandran A, Dharmaraj D, Snehlatha C, Viswanathan M (1992) Prevalence of glucose intolerance in Asian Indians: urban–rural difference and significance of upper body adiposity. Diabetes Care 15:1348–1355
11. Fagot-Campagna A et al (2000) Type 2 diabetes among North American children and adolescents: an epidemiologic review and a public health perspective. J Pediatr 136(5):664–672
12. Pinhas-Hamiel O (1996) Increased incidence of non-insulin-dependent diabetes mellitus among adolescents. J Pediatr 128(5 Pt 1):608–615
13. Bennett PH, Burch TA, Miller M (1971) Diabetes mellitus in American (Pima) Indians. Lancet 2:125–128
14. McKeigue PM, Miller GJ, Marmot MG (1989) Coronary heart disease in South Asians overseas—a review. J Clin Epidemiol 42:597–609
15. Abate N, Chandalia M (2007) Ethnicity, type 2 diabetes and migrant Asian Indians. Indian J Med Res 125:251–258
16. Haffner S, Temprosa M, Crandall J et al (2005) Intensive lifestyle intervention or metformin on inflammation and coagulation in participants with impaired glucose tolerance. Diabetes 54:1566–1572

17. Rosenwaike I, Hempstead K, Rogers RG (1987) Mortality differentials among persons born in Cuba, Mexico, and Puerto Rico residing in the United States, 1979–81. Am J Public Health 77:603–606
18. Østergård T et al (2006) Endothelial function and biochemical vascular markers in first-degree relatives of type 2 diabetic patients: the effect of exercise training. Metabolism 55(11):1508–1515
19. Klein R et al (1998) Relation of glycemic control to diabetic complications and health outcomes. Diabetes Care 21(3):c39–c43
20. Petrovic BD et al (2005) Gly482Ser polymorphism of the peroxisome proliferator-activated receptor-gamma coactivator-1 gene might be a risk factor for diabetic retinopathy in Slovene population (Caucasians) with type 2 diabetes and the Pro12Ala polymorphism of the PPARgamma gene is not. Diabetes Metab Res Rev 21(5):470–474
21. Altshuler David (2000) The common PPARγ Pro12Ala polymorphism is associated with decreased risk of type 2 diabetes. Nat Genet 26:76–80
22. Cheyssac C (2006) Analysis of common PTPN1 gene variants in type 2 diabetes, obesity and associated phenotypes in the French population. BMC Med Genet 7:44
23. Bento JL (2004) Association of protein tyrosine phosphatase 1B gene polymorphisms with type 2 diabetes. Diabetes 53:3007–3012
24. Wang RQ (2011) ENPP1/PC-1 gene K121Q polymorphism is associated with obesity in European adult populations: evidence from a meta-analysis involving 24324 subject. Biomed Environ Sci 24(2):200–206
25. Fu D, Cong X, Ma Y et al (2013) Genetic polymorphism of glucokinase on the risk of type 2 diabetes and impaired glucose regulation: evidence based on 298, 468 subjects. PloS ONE 8(2):e55727
26. Florez JC, Jablonski KA, Kahn SE et al (2007) Type 2 diabetes-associated missense polymorphisms KCNJ11 E23K and ABCC8 A1369S influence progression to diabetes and response to interventions in the diabetes prevention program. Diabetes 56:531–536
27. Sakamoto Y et al (2007) SNPs in the KCNJ11-ABCC8 gene locus are associated with type 2 diabetes and blood pressure levels in the Japanese population. J Hum Genet 52:781–793
28. Glyon AL (2003) Large-scale association studies of variants in genes encoding the pancreatic beta-cell KATP channel subunits Kir6.2 (KCNJ11) and SUR1 (ABCC8) confirm that the KCNJ11 E23K variant is associated with type 2 diabetes. Diabetes 52(2):568–572
29. Horikawa Y (2000) Genetic variation in the gene encoding calpain-10 is associated with type 2 diabetes mellitus. Nat Genet 26:163–175
30. Gloyn AL, Braun M, Rorsman P (2009) Type 2 diabetes susceptibility gene TCF7L2 and its role in beta-cell function. Diabetes 58:800–802

Chapter 6
Biohardening of Micropropagated Plants with PGPR and Endophytic Bacteria Enhances the Protein Content

Sunitha Panigrahi, K. Aruna Lakshmi, Y. Venkateshwarulu and Nikkita Umesh

Abstract The ability to grow plant tissues in vitro and to control their development forms the basis of many practical applications in agriculture, horticulture and industrial chemistry. Establishment of a cell, tissue or organ culture and regeneration of plantlets under in vitro conditions has opened up new avenues in the area of plant biotechnology. Micropropagation is one such method where the plants are grown under in vitro conditions to provide a better survival rate and mass propagation in a short time, it is achieved through the establishment of the explants and their initial growth in vitro, followed by transferring them onto the field or in a greenhouse. This has triggered a great interest in understanding plant physiology and performing a comprehensive study regarding the change in physiology and morphology under influence of various useful bacterial combinations. In this review, plant growth promoting rizobacteria (PGPR) were used in the bio priming of the micro propagated plants (banana), these plants showed increased growth of the root length; shoot length, internode diameter and number of leaves. All parts of the banana plant have medicinal applications: the flowers in bronchitis and dysentery and on ulcers; cooked flowers are given to diabetics; the astringent plant sap in cases of hysteria, epilepsy, leprosy, fevers, hemorrhages, acute dysentery and diarrhea, and it is applied on hemorrhoids, insect and other stings and bites; young leaves are placed as poultices on burns and other skin afflictions;

S. Panigrahi (✉)
Department of Life Science, St. Mary's College, Hyderabad, India
e-mail: sritha17@yahoo.com

K. Aruna Lakshmi
Department of Biotechnology, GITAM University, Visakhapatnam, India
e-mail: ravikommaraju@gmail.com

Y. Venkateshwarulu
Department of Microbiology, Andhra University, Visakhapatnam, India
e-mail: raovyechuri@yahoo.co.in

N. Umesh
St. Mary's College, Hyderabad, India
e-mail: nikkitha.umesh@gmail.com

A. Kumar, *Biotechnology and Bioforensics*, Forensic and Medical Bioinformatics,
DOI: 10.1007/978-981-287-050-6_6,

the astringent ashes of the unripe peel and of the leaves are taken in dysentery and diarrhea and used for treating malignant ulcers; the roots are administered in digestive disorders, dysentery and other ailments; banana seed mucilage is given in cases of diarrhea in India. The changes in protein expression as a consequence of environmental stimuli were evaluated which led to the conclusion that the bio hardened plants showed high protein content but had the similar genetic content when compared against the control plants. Thus increase in the protein content enhances these medicinal qualities of banana plant. Further, a fidelity test was conducted on the bio primed plants to check the genetic variations.

Keywords Micropropagation · In vitro culture · PGPR · Biopriming

6.1 Introduction

Micropropagation has now become a multibillion dollar industry, practised all over the world it's in fact a complex multistep process involving numerous different starting tissues and cell types [1, 2]. The commercial multiplication of a large number of diverse plant species represents one of the major success stories of utilizing tissue culture technology. The setback in micro propagation is that most of the plants are not resistant to the ex-vitro conditions. These plants do not have full dermal coverings and are very fragile when first brought onto the field. They are rapidly exposed to various diseases and harmful microbes that reduce their sustenance level. To make this right, bacterial inoculation at the beginning of the acclimatisation phase must be performed. Biopriming (or) biohardening refers to the treatment of tissue cultured banana plantlets with microbial inoculants to strengthen the plantlets against biotic and abiotic stresses [3]. *Musa acuminata* (Banana) is one of the most important fruit crops of world as well as of India. Banana could be considered poor man's apple, and it is the cheapest among all other fruits in the country [4]. In recent years, micropropagation of banana has gained attention because of its potential to provide genetically uniform, pest and disease-free planting material [5]. Del Carmen Jaizme-Vega et al. [6] evaluated the effect of a rhizobacteria consortium of *Bacillus* spp. on the first developmental stages of two micropropagated bananas. The morphological and physiological characters were also well pronounced in the treated plants indicating its plant growth promoting nature [3]. It has been proved that treating plants with certain symbiotic soil microorganisms help the plant in better absorption of nutrients from the surrounding environment, improves the drought resistance and disease resistance of the plant thereby increasing the survival rate of propagated plants [7]. There is a need for different techniques that increase the input efficiency, and plant growth promoting rhizobacteria (PGPR) have proved to be a major tool. PGPR can affect on plant growth by production and release of secondary metabolites, lessening or preventing deleterious effects of phytopathogenic organisms in the rhizosphere and/or phyllosphere, and/or facilitating the availability and uptake of certain nutrients like N, P, and Fe from the root environment [8].

The current work was carried out in continuation to our previous research on hardening of the Micro propagated plants using micro organisms (PGPR) as biological agents. We used single, dual, triple, four, five and six microbial consortium and prepared the standard solutions into which the roots of the plantlets were dipped and incubated at the time of Green house stage of the tissue culture. The plants produced in this way gave very good results in terms of their Root, Shoot, Leaf Primordia, No of leaves and macro and micro nutritive value. The present research highlighted the presence of high levels of proteins in the PGPR inoculated plants which enhanced the growth of the plants when compared against the control plants. These proteins were then extracted, estimated and the concentrations of these proteins were compared to know which plantlet produced the highest yield of the enzymes. Later, In order to verify the genetic stability the Granine variety of the banana which was treated with PGPR was tested for the genetic fidelity test.

6.2 Materials and Methods

6.2.1 Hardening Process

The normal treatment given to the tissue cultured plants during the hardening process is, the plantlets are first transferred to distilled water in order to avoid dehydration, then they are sprayed with some fungicides and then transferred to the vermiculate or soil peat, and perlite containing N:P:K at a ratio of 18:6:12. Then they are grown in green house where the fungicides are sprayed at regular intervals to avoid contamination. Humid conditions and ambient temperature are maintained for a short period of time after which the intensity of light and temperature are increased gradually. Plants are later transferred to the field. This was the normal procedure which is adopted by the AG Biotech plants. But in this research, the above hardening procedure was manipulated to get the desired results. Plantlets were not sprayed with any fungicide when they were removed from the MS media and the soil which was used for planting purpose was sterile, i.e., devoid of any microorganisms.

6.3 Isolation of Bacteria from Soil

The micro organisms which were used in this study are isolated and enumerated from the soils of vegetative fields of Andhra Pradesh, India. There were seven different soil samples taken from various districts. Later the microorganisms were serial diluted, identified using various techniques, and then enumerated.

6.4 Inoculation of the Bacteria

The bacteria were inoculated to the roots of the micropropagated plants.

The Microorganisms were dissolved in the sterile distilled water and were treated to the plants in various combinations, 0.5, 1.0, 1.5, 2.0, 2.5, 3.0 %. When the bacteria was inoculated to the tissue cultured plants, the root length, shoot length, no of leaves and the leaf primordial of the plantlets were observed and documented [9].

The micro organisms were treated to the plants in single parameter, dual, three, four, five combinations, and the last was the consortium, each parameter had ten plantlets compared against control, the consortium with 2.5 % was taken as the optimum value for the plants.

6.5 Plant Collection

Four different categories of plants were taken such as the control (T1), Plants treated with PGPR (T2), the AG Biotech treated plants (T3) and the AG Biotech plants treated with PGPR (T4).

6.6 Nutrient Analysis

The macro and micro nutrient analysis was determined by Perkin-Elmer Analyst, 300 single beam Atomic Absorption Spectrophotometer and the data was obtained and documented [10].

6.7 Quantitative Estimation of Enzyme by Lowry's Method

The plant extract of the T1, T2, T3, T4 samples were taken and the protein concentration present in the plant samples was estimated against the control, the protocol used is as per Lowry's method. Protein estimation was done with Folin-Phenol reagent using BSA as a standard. Folin's reagent transforms the BSA solution from colorless to purple. After the addition of alkaline solution, all the contents are incubated at room temperature for 10 min. Then folin's reagent is added and again incubated at room temperature for 30 min. The principle behind the Lowry method of determining protein concentrations lies in the reactivity of the peptide nitrogen[s] with the copper [II] ions under alkaline conditions and the subsequent reduction of the Folin-Ciocalteau phosphomolybdic phosphotungstic

Table 6.1 Soil samples collected from various places of AP

Sample no.	Crop field/Grounds	Place
1	Paddy field	Manoharabad, Rangareddy district
2	Vegetable field	Anakapally, Vizag district
3	Ground nut field	Nandyal, Kurnool district
4	Sun flower field	Miriyalaguda, Nalgonda district
5	Red gram field	Eluru, East Godavari district
6	Pearl Millet field	Sngareddy, Medak district
7	Sesame field	Aurmoor Nizamabad district
8	Sea coast	Vizag. Vishakhapatnam district
9	Dump site	Hussain Sagar Lake, Hyderabad district

acid to heteropoly molybdenum blue by the copper-catalyzed oxidation of aromatic acids. The color change is determined calorimetrically at 540 nm. The Lowry method is sensitive to pH changes and therefore the pH of assay solution should be maintained at 10–10.5. This step would finally confirm the amount of protein content present among the test samples.

6.8 Fidelity Test

As there was an increase in the growth rates, nutrient rate for the biohardened plants, to ensure the genetic composition of the bacteria was similar to the parental true type there was a genetic mapping been conducted and the results were documented [11].

6.9 Results and Discussion

6.9.1 Enumeration of the Microorganisms

There were six bacteria which were enumerated in the soils of Andhra Pradesh, they are Rhizobium, Acetobacter, VAM, Pseudomonas, Azospirullum, PSB (Table 6.1).

Each bacterium showed varied colony forming units, the actual number of CFU was analyzed for the inoculation considering the volume of loop to be 0.01 ml.

The viable counts ranged in the soil samples were *Rhizobium* (2.41×10^4 CFU/ml), *Pseudomonas* (1.42×10^3 CFU/ml), PSB (1.5×10^6 CFU/ml), *Azospirillum* (1.82×10^5 CFU/ml), *Acetobacter* (2.0×10^7 CFU/ml).

Conformation of the Bacterial samples was analyzed by performing grams staining, IMVIC tests [9] (Table 6.2).

Table 6.2 Confirmative test for soil bacteria

Test	*Rhizobium*	*Pseudomonas*	PSB	*Acetobacter*	*Azospirillum*
Gram's staining	−ve	−ve	+ve	−ve	−ve
Methyl red	+ve	−ve	+ve	+ve	+ve
Voges-Proskaeur	−ve	−ve	+ve	−ve	+ve
Indole	−ve	−ve	−ve	−ve	+ve
Citrate	−ve	+ve	+ve	+ve	−ve
Catalase	+ve	+ve	+ve	+ve	+ve
Oxidase	+ve	+ve	+ve	+ve	+ve

Table 6.3 Showing the macro nutrient content of the plants

Plants	N %	P %	K %
T1	1.43	0.09	1.74
T2	2.34	0.11	2.33
T3	1.89	0.16	1.88
T4	2.76	0.27	3.36

Table 6.4 Showing the micro nutrient content of the plants

Plants	Mn %	Ca %	Zn %	Mg %	Fe %
T1	1.463	0.391	6.21	0.08	0.96
T2	1.789	0.14	9.27	0.16	1.43
T3	1.532	0.571	10.07	0.21	1.061
T4	2.035	0.768	14.8	0.32	1.721

6.9.2 Plant Isolation

When there was Single, dual, triple, four, five and the consortium of the bacterial combinations were inoculated to the tissue cultured plants, the root length, shoot length, no of leaves and the leaf primordial of the plantlets were observed and documented [9].

The Mean concentration of macro nutrients is in the order of (highest to lowest): K (1.74) > N (1.43) > P (0.09) was recorded. Table 6.3 shows that the nitrogen content was less in the plants which were treated with the Consortium and slightly lower with a difference of 0.43 % in the plants from the AG Biotech laboratory treated plants along with the consortium, Phosphorous content is less with a difference of 0.16 %, and potassium with 1.03 %.

The Mean concentration of micro nutrients is in the order of (highest to lowest): Zn (6.21) > Mn (1.463) > Fe (0.96) > Ca (0.391) > Mg (0.08) as shown in Table 6.4.

Table 6.5 Comparative analysis of the quantitative protein

Plant samples	Volume of protein (protease)	DW (ml)	Vol of reagent copper sulphate (ml)[a]	Folin Ciocalteau reagent (ml)[b]	O.D. 540 nm
Blank	0	1	2	1	0
T1(20s)	0.3	0.2	2	1	0.02
T2(30s)	0.3	0.2	2	1	0.48
T3(35s)	0.3	0.2	2	1	0.37
T4(40s)	0.3	0.2	2	1	0.52

[a] Incubation at room temperature for 30 min
[b] Incubation at room temperature for 10 min

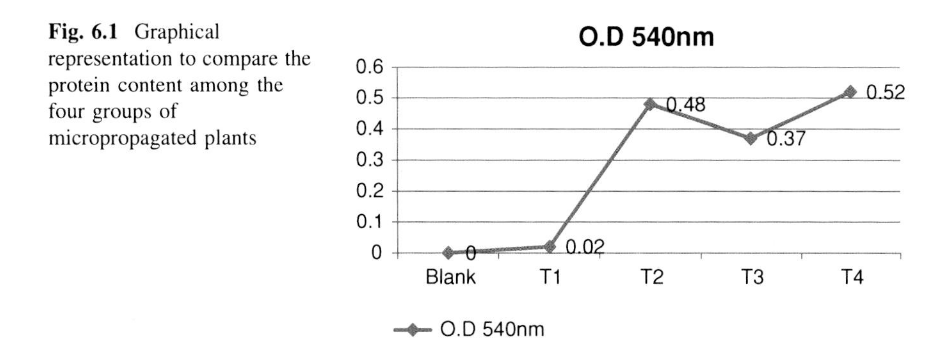

Fig. 6.1 Graphical representation to compare the protein content among the four groups of micropropagated plants

Table 6.4 shows the Manganese content which was less in the plants which were treated with the Consortium and slightly lower with a difference of 0.246 % in the plants which were from the AG Biotech laboratory.

Treated plants along with the consortium, Calcium content are less with a difference of 0.628 %, Zinc with 5.53 %, Magnesium with 0.16 % and Iron with 0.291 %. The foliar macro and macro nutrient content which was organically grown in Canary Island [12].

6.10 Protein Concentration

6.10.1 Comparative Analysis of the Quantitative Protein

The concentration of protein thus produced form the above test cultures is compared so as to propose one standard treatment to the initial inoculums that can produce high yields of the protein (Table 6.5 and Fig. 6.1).

The concentration of proteins followed the pattern of T4 > T2 > T3 > T1, which means that the biologically hardened plants showed more amount of proteins, than any other plants. The plant treated with a consortium of the bacteria showed the highest value of protein content.

Plants treated with mixtures of rhizobacterial and endophytic bacterial formulations under field conditions recording 33.33 % infection with 60 % reduction over control. Transcripts encoding endochitinase, beta-1,3-glucanase, a thaumatin-like protein, ascorbate peroxidase, metallothionein, and a putative senescence-related protein increased early in ripening [13].

Banana (*Musa acuminata*, cv Dwarf Cavendish) proteins were extracted from pulp tissue at different stages of ripening and analyzed by two-dimensional electrophoresis. The results provide evidence of differential protein accumulation during ripening. Two sets of polypeptides have been detected that increase substantially in ripe fruit [14].

Inoculation of rhizobacteria was showed beneficial to the banana plantlet in saline conditions through increment of growth and improvement in rooting system. Thus, these bacterial strains could be used as a bioenhancer for growth of in vitro banana plantlets [7].

6.11 Conclusion

In conclusion we can establish that when a suitable PGPR (Biohardening treatment) given to the micropropagated plants without any usage of fungicides at a particular concentration of 2.5 % increases the micro, macro nutrient values, and a constant increase in the protein content boosted the growth of the micropropagated plants. This infers that the growth of the plant will be more and the yield of the plant will increase.

References

1. Rani V, Raina SN (2000) Genetic fidelity of organized meristem-derived micropropagated plants: a critical reappraisal. Vitro Cell Dev Biol Plant 36(5):319–330
2. Aitken-Christie J, Connett M (1992) Micropropagation of forest trees. In: Kurata K, Kozai T (eds) Transplant production systems. Kluwer, Dordrecht, pp 163–194
3. Harish S, Kavino M, Kumar N, Saravanakumar D, Soorianathasundaram K, Samiyappan R (2008) Biohardening with plant growth promoting rhizosphere and endophytic bacteria induces systemic resistance against banana bunchy top virus. Appl Soil Ecol 39(2):187–200
4. Thakker JN, Patel S, Dhandhukia PC (2011) Induction of defense-related enzymesin susceptible variety of banana: role of Fusarium derived elicitors. Achieves Phytopathol Plant Prot 44(20):1976–1984
5. Ayyadurai N, Ravindra Naik P, Sreehari Rao M, Sunish Kumar R, Samrat SK, Manohar M, Sakthivel N (2006) Isolation and characterization of a novel banana rhizosphere bacterium as fungal antagonist and microbial adjuvant in micropropagation of banana. J Appl Microbiol 100(5):926–937
6. del Carmen Jaizme-Vega M, Rodríguez-Romero AS, Guerra MSP (2004) Potential use of rhizobacteria from the *Bacillus* genus to stimulate the plant growth of micropropagated bananas. Cambridge J 59(02):83–90

7. Mahmood M, Rahman ZA, Saud HM, Shamsuddin ZH, Subramaniam S (2010) Influence of rhizobacterial and agrobacterial inoculation on selected physiological and biochemical changes of banana cultivar, Berangan (AAA) plantlets. J Agric Sci 2(1)
8. Esitken A (2011) Use of plant growth promoting rhizobacteria in horticultural crops, Bacteria Agrobiology Crop Ecosyst 189–235
9. Panigrahi S, Aruna Lakshmi K, Bathina S (2013) A biological approach to harden the micropropagated plants using the soil Microorganisms. Micro Macro Nutr Anal Helix 3:324–327
10. Panigrahi S, Aruna Lakshmi K, Mir N (2013) Micropropagation and plant strengthening of tissue cultured plants, inoculated with several bacterial strains. IJSR 2(8):15–17
11. Panigrahi S, Aruna Lakshmi K (2013) Fidelity testing, an approach to ensure the genomic stability of the biologically hardened micropropagated plants. IJSER 7:145–149
12. Alvarez CE, Ortega A, Fernández M, Borges AA (2001) Growth, yield and leaf nutrient content of organically grown banana plants in the Canary islands. Fruits 56:17–26
13. Clendennen SK, May GD (1997) Differential gene expression in ripening banana fruit. Plant Physiol 115(2):463–469
14. Dominguez-Puigjaner E, Vendrell M, Ludevid DM (1992) Differential protein accumulation in banana fruit during ripening. Plant Physiol 98(1):157–162

Chapter 7
Effect of Plant Growth Regulators on Morphological, Physiological and Biochemical Parameters of Soybean (*Glycine max* L. Merrill)

R. Ramesh and E. Ramprasad

Abstract Plant growth and developmental processes are very much regulated by certain chemical substances called Growth regulators. Growth regulators are known to improve the physiological efficiency including photosynthetic ability and can enhance the effective partitioning of accumulates from source and sink in the field crops. The present study was conceptualized and executed with the prime objective of study the effect of chlormequat chloride, NAA, Mepiquat chloride and Brassinosteroids on morphological, physiological and biochemical parameters of soybean. The field trial was conducted following randomized block design with nine treatments replicated thrice. The basic material for the present investigation consisted of soybean cv. Js-335 and two growth promoting (NAA and Brassinosteroid) and growth retarding substances (chlormequat chloride and mepiquat chloride). These growth regulators were sprayed at flower initiation stage. The Morphophysiological parameters, namely, Plant height, number of branches, number of trifoliates per plant, dry matter accumulation in leaf, stem and reproductive parts, LAI, CGR and RGR was observed to increase significantly with the application of NAA (20 ppm) and brassinosteroid (25 ppm). However, it decreased with the application of chlormequat chloride and mepiquat chloride. The Biochemical parameters, namely, chlorophyll content was observed to increase significantly with application of NAA (20 ppm), brassinosteroid (25 ppm), mepiquat chloride 5 %, AS (5 %) and chlormequat chloride 50 % SL at different concentrations compared to control and water spray but whereas fluorescence emission and photosynthetic rate were noticed to be non-significant. A significant increase in the seed protein content was also noticed with the application of NAA (20 ppm), brassinosteroid (25 ppm), mepiquat chloride 5 %, AS (5 %) and

R. Ramesh (✉)
Department of Crop Physiology, Acharya N. G. Ranga Agricultural University, Hyderabad, India
e-mail: rameshpphy@gmail.com

E. Ramprasad
Department of Plant Molecular Biology and Biotechnology, Acharya N. G. Ranga Agricultural University, Hyderabad, India

A. Kumar, *Biotechnology and Bioforensics*, Forensic and Medical Bioinformatics,
DOI: 10.1007/978-981-287-050-6_7,

chlormequat chloride at different concentrations, compared to control and water spray. In conclusion, the study revealed the superiority of NAA (20 ppm) treatment for majority of the morphological, physiological and biochemical parameters at different growth stages, compared to other growth regulator and control treatments studied in the present investigation for Rabi soybean.

Keywords Growth regulators · Randomized block design · Soybean cv. Js-335 · Morphophysiological parameters · Biochemical parameters

7.1 Introduction

Soybean (*Glycine max.* L. Merril) is a highly nutritive and energy rich dual purpose rainy season monocarpic legume crop with biologically effective proteins (43 %), edible oil (20 %), vitamins, minerals, salts and essential amino acids. Because of its versatility, soybean is popularly known as "Miracle Bean" and is being exploited in many agro-based industries with innumerable ways. India occupies an area of 8.89 m ha with a production of 10.96 mt and productivity of 1,235 kg ha^{-1} (Centre for Monitoring Indian Economy 2012). The evaluation of morpho-physiological and biochemical traits of crops namely, Plant height, number of branches, number of leaves per plant, dry matter accumulation in leaf, stem and reproductive parts, LAI, CGR and RGR, fluorescence, photosynthetic rate, chlorophyll content, germination, flowering, pod and seed development in addition to senescence either by reducing (growth retardant) or enhancing plant growth (growth promoters) indicates crop growth patterns which are reflected in final yield and thus, influences crop productivity. Plant growth regulators have been reported to be an effective tool for increasing crop yields due to their important role in various physiological and biochemical processes in plant, leading to rapid change in phenotype of the plant within the season to achieve desirable results. The use of growth regulators has been gaining more importance in the recent years for improvement of crop yield potential and quality of produce. In this context, there is an urgent need to identify suitable growth regulators for improving yield potential by changing the various above mentioned parameters in soybean.

7.2 Materials and Methods

The investigation was conducted on 'cv. JS-335' soybean genotype during rabi with nine treatments replicated thrice involving exogenous application of PGRs [NAA (20 ppm), brassinosteroid (25 ppm), mepiquat chloride 5 %, AS (5 %) and chlormequat chloride at different concentrations and a control i.e., no spray] in a randomized block design in the College Farm, College of Agriculture, Rajendranagar, Hyderabad. Foliar application of PGRs was made at flower initiation stage. Plant based observations viz., namely, Plant height, number of branches, number of trifoliates per plant, dry matter accumulation in leaf, stem and reproductive parts (60 and 75 days up to harvest), LAI, CGR and RGR were recorded on five randomly selected and tagged plants at 15 days intervals from 15 DAE (days after emergence) till up to 75 DAE and at the time of harvest. The leaf area (cm^2) was measured using LI-3100 Leaf Area Meter (LICOR-Lincoln, Nebraska, USA) leaf area index (LAI), Crop growth rate ($g\ m^{-2}\ d^{-1}$), Relative growth rate ($g\ g^{-1}\ d^{-1}$) were estimated. Afterwards, these leaf samples were subsequently dried hot-air oven at 60–70 °C. The dried samples were then weighed to record data on dry matter production. The values were expressed as $g\ m^{-2}$. Biochemical characters like chlorophyll content was measured using SPAD-502 (Soil Plant Analytical Development) meter, Leaf fluorescence was measured using fluorescence induction monitor (FIM-1500), Measurement of seed protein content was estimated by powdered the dried seed material. Thereafter the nitrogen percentage in the seed was estimated by taking 0.1 g of powdered seed sample following the kjeldahl procedure as given in [1]. Protein percentage was then estimated by multiplying the nitrogen percentage with the factor, 6.25. Then Nitrogen harvest index (%) was measured. Finally the data were analysed statistically following the method of [2].

7.3 Results and Discussion

7.3.1 Morphological Traits

The data on plant height at different growth stages (Table 7.1) indicated that plant height increased with crop age. The growth promoters, especially NAA and brassinosteroid increased plant height, compared to water spray, control and the growth retardants studied in the present investigation. A comparison of the growth promoter treatments further indicated that NAA at 20 ppm resulted in maximum plant height, compared to other treatments studied in the present investigation. The increased plant height may be due to the stimulating action of auxin which softens the cell wall by increasing its plasticity or may be the oxidative decarboxylation of synthetic auxins which could not be catalyzed by the enzyme peroxidase. Similar results have also been reported earlier [3] in black gram due to application of NAA. The application of brassinosteroid (25 ppm), another growth promoter studied in the present investigation had also resulted in increased plant height, on par with NAA (20 ppm) and significantly greater, compared to control and water spray. Similar results in increase in plant height was observed [4] due to application of brassinosteroid and attributed it to increased photosynthetic rate in soybean. The growth retardants studied in the present investigation, namely, chlormequat chloride and mepiquat chloride recorded a decrease in plant height,

Table 1 Effect of different growth regulators on plant height (cm), no. of branches/plant and no. of trifoliates/plant in Rabi soybean

Treatments		Plant height (cm)						No. of branches						No. of trifoliates					
		Days after emergence					At harvest	Days after emergence					At harvest	Days after emergence					At harvest
		15	30	45	60	75		15	30	45	60	75		15	30	45	60	75	
T1	Chlormequat chloride 50 % SL (137.5 g a.i/ha)	10.14	14.46	20.15	25.95	26.07	26.16	2.99	3.65	4.05	4.41	4.94	4.94	3.9	11.16	19.59	22.19	20.42	18.6
T2	Chlormequat chloride 50 % SL (162.5 g a.i/ha)	9.64	14.37	20.08	24.06	24.52	24.65	2.96	4.57	4.91	5.02	5.39	5.39	3.99	10.54	19.44	21.61	20.57	13.08
T3	Chlormequat chloride 50 % SL (187.5 g a.i/ha)	8.76	11.87	17.21	19.83	20.43	20.46	2.99	4.09	4.86	4.97	5.28	5.28	4.11	11.68	19.34	22.08	20.84	14.77
T4	Chlormequat chloride 50 % SL (375 g a.i/ha)	8.39	9.27	14.92	19.15	20.04	20.08	2.96	3.38	3.84	4.17	4.18	4.18	3.82	9.73	19.15	19.71	18.37	16.07
T5	Alpha naphthyl acetic acid (NAA) 20 ppm	11.41	23.46	28.54	31.79	32.25	32.28	2.99	4.03	5.07	5.19	5.51	5.51	5.02	14.12	23.47	30.62	27.27	22.95
T6	Mepiuat chloride 5 % AS (5 %)	8.48	12.01	16.14	20.72	22.07	22.09	2.99	4.19	4.57	4.75	5.06	5.06	4.14	12.02	19.64	22.33	22.34	16.66
T7	Brassinosteroid (25 ppm)	10.87	14.69	27.69	30.88	31.03	31.12	2.86	3.72	4.02	4.19	4.24	4.24	4.7	13.3	21.16	23.43	22.69	20.19
T8	Water	10.39	13.73	21.32	28.07	28.79	28.82	2.86	3.62	3.77	4.08	4.16	4.16	3.33	9.28	18.34	19.54	19.23	12.02
T9	Control	10.22	14.72	20.44	26.53	26.99	27.06	2.83	3.28	3.42	3.76	3.9	3.9	3.2	8.13	16.75	18.9	17.27	10.92
	Mean	9.81	14.29	20.72	25.22	25.8	25.86	2.94	3.84	4.28	4.5	4.74	4.74	4.02	11.11	19.65	22.27	21	16.14
	Sed	0.21	1.16	0.5	0.86	0.74	0.77	0.04	0.09	0.11	0.15	0.12	0.12	0.03	0.37	0.41	0.93	1.38	0.38
	CD (0.05)	NS	NS	1.07	1.84	1.58	1.65	NS	NS	0.22	0.31	0.24	0.24	NS	0.78	0.87	1.96	2.93	NS

compared to water spray, control and the growth promoters. A similar reduction in the plant height of soybean crop with the application of chlormequat chloride (40 ppm) at 35 DAE, compared to NAA and GA was reported by Govindan et al. [5]. The mechanism of reduction in plant height appears to be due to reduction in cell division and its expansion.

The number of branches per plant is an important morphological character, directly related to yield in soybean. In the present investigation, number of branches per plant increased gradually up to 75 days (Table 7.1). Further, the different growth regulator treatments studied differed significantly at 45, 60, 75 DAE and harvest and maximum number of branches per plant were recorded in NAA (20 ppm), followed by chlormequat chloride 50 % SL (162.5 a.i/ha). Similar results were reported by Deotale et al. [6] in soybean. Application of mepiquat chloride 5 % AS (5 %) and brassinosteroid (25 ppm) had also recorded significantly higher number of branches per plant, compared to control in the present study. Similar results were reported earlier in black gram [8] and green gram [7].

In general, leaf is considered as an important functional unit of plant which contributes to the formation of yield. The number of trifoliate leaves increased gradually from 15D AS to 60 DAE, and thereafter declined (Table 7.1) due to senescence. Significant influence of the growth regulator treatments was noticed at 30, 45, 60, 75 DAE and harvest. All treatments, with the exception of chlormequat chloride 50 % SL (375 a.i/ha) at 60 and 75 DAE, increased the number of trifoliates significantly, compared to control. The positive influence of chlormequat chloride and NAA [8]; and mepiquat chloride and brassinosteroid [9] on number of trifoliates per plant were also reported earlier. Among the growth regulators, NAA (20 ppm) was more effective followed by brassinosteroid (25 ppm).

Poor partitioning of photo assimilates to the growing reproductive parts is one of the major constraints in pulses. This can be overcome by applying synthetic plant growth regulators. The dry matter production in leaves increased up to 60 DAE and declined (Table 7.2) thereafter till harvest in all the growth regulator treatments including control. Among the treatments, the leaf dry weight was significantly higher with the application of NAA (20 ppm), brassinosteroid (25 ppm), mepiquat chloride 5 % AS (5 %) and chlormequat chloride 50 % SL (137.5 and 162.5 a.i/ha), compared to control and water spray, due to the beneficial effect of these growth regulators on leaf development. Similar results with foliar application of 50 ppm NAA at flowering in soybean and pigeon pea were obtained by Merlo et al. [10]. The data recorded (Table 7.2) on dry matter production in stem revealed gradual increase up to 75 DAE. Among different growth regulator treatments NAA (20 ppm) exhibited maximum stem dry matter followed by brassinosteroid (25 ppm), chlormequat chloride 50 % SL (137.5 a.i/ha) over control and other treatments studied. Similar results in increase in stem dry matter was observed by Chandrasekar and Bangarusamy [11] in green gram with the application of NAA @ 40 ppm. Dry matter production, particularly in reproductive parts is an important yield contributing character. There was a gradual

Table 2 Effect of different growth regulators on leaf dry weight (g m^{-2}), stem dry weight (g m^{-2}) and reproductive parts (g m^{-2}) in Rabi soybean

		Leaf dry weight						Stem dry weight						Reproductive parts		
		Days after emergence					At harvest	Days after emergence					At harvest	Days after emergence		At harvest
		15	30	45	60	75		15	30	45	60	75		60	75	
T_1	Chlormequat chloride 50 % SL (137.5 a.i/ha)	4.03	20.63	116.21	144.23	112.06	67.26	2.38	16.30	105.23	141.53	178.11	130.45	259.28	320.20	422.02
T_2	Chlormequat chloride 50 % SL (162.5 a.i/ha)	4.10	23.06	111.40	140.08	101.54	97.07	2.99	17.50	103.07	138.66	172.83	127.80	289.42	363.16	427.77
T_3	Chlormequat chloride 50 % SL (187.5 a.i/ha)	4.64	24.60	98.40	132.54	80.96	37.03	3.76	20.13	93.22	135.29	155.28	122.99	218.21	310.23	415.54
T_4	Chlormequat chloride 50 % SL 375 (a.i/ha)	4.21	23.53	90.60	117.28	51.69	29.34	3.31	20.01	70.16	120.54	132.30	97.33	201.11	305.18	366.44
T_5	Alpha naphthyl acetic acid (NAA) 20 ppm	4.07	23.24	125.50	165.47	143.10	127.38	2.71	17.17	109.41	153.48	185.10	169.98	300.11	427.74	488.30
T_6	Mepiuat chloride 5 % AS (5 %)	4.13	23.34	115.41	143.37	121.79	107.21	3.05	16.71	94.26	136.52	167.70	124.93	288.64	340.15	425.16
T_7	Brassinosteroid (25 ppm)	3.73	20.29	124.30	163.60	142.36	107.27	2.70	16.45	108.06	149.27	182.97	157.65	290.94	345.19	435.20
T_8	Water	3.58	21.54	84.63	106.67	62.42	58.12	2.30	16.02	78.71	134.54	139.35	108.34	158.11	237.31	276.14
T_9	Control	3.48	19.24	77.30	101.94	57.47	47.81	2.20	14.51	73.97	126.17	131.30	101.59	155.27	233.25	274.55
	Mean	**4.00**	**22.16**	**104.86**	**135.02**	**97.04**	**75.39**	**2.82**	**17.20**	**92.90**	**137.33**	**160.55**	**126.78**	**240.12**	**320.27**	**392.35**
	Sed	1.72	0.96	1.02	1.62	2.15	1.46	0.48	0.69	1.51	2.41	2.12	1.09	2.15	3.60	4.33
	C.D (0.05)	NS	NS	2.18	3.46	4.57	3.12	NS	NS	3.22	5.13	4.52	3.03	4.59	7.67	9.23

increase in dry matter production of pods (Table 7.2) from 60 DAE to harvest stage, and the highest dry matter was observed at harvest stage. Further, among the growth regulators treatments, NAA (20 ppm) followed by brassinosteroid (25 ppm) and chlormequat chloride 50 % SL (162.5 a.i/ha) had recorded maximum dry matter in pods, compared to control and other treatments studied. Similar increase in dry matter of reproductive parts with the application of NAA @ 20 ppm in soybean [4]; and chlormequat chloride in soybean [12] were reported earlier.

7.3.2 Physiological Traits

Leaf area index is considered to be one of the photosynthetic determinants in crop plants and in the present study, it increased gradually (Table 7.3) from 30 to 75 DAE and decreased thereafter due to senescence and ageing of leaves. Application of growth promoters, namely, NAA (20 ppm) and brassinosteroid (20 ppm) had resulted in higher LAI, compared to the growth retardants, mepiquat chloride and chlormequat chloride, in addition to water spray and control, at 30, 45, 60 and 75 DAE, probably due to their positive effect on cell division and cell elongation leading to enhanced leaf growth. Further, the application of growth retardants, namely, mepiquat chloride 5 % AS (5 %) and chlormequat chloride 50 % SL at different concentrations had also resulted in higher LAI, compared to control, in the present investigation. The positive influence of mepiquat chloride [7] and chlormequat chloride [13, 14] has also been reported earlier.

The average daily increment in biomass production, namely, CGR is an important useful tool for estimating production efficiency, enabling comparison between the treatments. The observations recorded (Table 7.3) on CGR in the present study revealed that most of the CGR values were maximum at 45–60 DAE. There was a gradual increase in CGR values from 15–30 DAS to 45–60 DAE, and thereafter it declined. Application of NAA (20 ppm), followed by brassinosteroid (25 ppm) had recorded high CGR over control and other treatments studied.

RGR, an efficiency index, declined with the advancement of crop growth from 15–30 DAE to 75DAE-harvest, due to a decline in crop growth rate (CGR) and more particularly, in the rate of dry matter production. Maximum RGR values (Table 7.3) were attained at 15–30 DAE. The application of NAA (20 ppm) had resulted in significantly high RGR at 30–45, 60–75 and 75-harvest growth stages. Further, Mepiquat chloride 5 % AS at 75-harvest and chlormequat chloride 50 % SL applied at 137.5 a.i/ha at 30–45 DAE; and at 162.5 and 187.5 a.i/ha at 75-harvest had also recorded high RGR, compared to control. The findings are in conformity with the reports of [15] in soybean.

Table 3 Effect of different growth regulators on leaf area index, crop growth rate (CGR) (g m^{-2} d^{-1}), relative growth rate (*RGR*) (g g^{-1} d^{-1}) in Rabi soybean

	Treatments	LAI						CGR					RGR				
		Days after emergence					At harvest	Days after emergence					Days after emergence				
		15	30	45	60	75		15–30	30–45	45–60	60–75	75-Harvest	15–30	30–45	45–60	60–75	75-Harvest
T1	Chlormequat chloride 50 % SL (137.5 g a.i/ha)	0.14	0.75	2.4	4.52	3.43	2.34	2.03	12.3	21.57	4.36	0.62	0.1167	0.1194	0.06	0.0075	0.001
T2	Chlormequat chloride 50 % SL (162.5 g a.i/ha)	0.13	0.73	2.23	4.32	3.07	2.23	2.23	11.59	23.58	4.62	1.01	0.1163	0.111	0.0649	0.0077	0.0016
T3	Chlormequat chloride 50 % SL (187.5 g a.i/ha)	0.16	0.72	2.13	4.26	3.05	2.15	2.42	9.79	19.63	4.03	1.94	0.1115	0.097	0.0621	0.0078	0.0035
T4	Chlormequat chloride 50 % SL (375 g a.i/ha)	0.14	0.71	2.06	4.18	2.91	2.04	2.4	7.81	18.54	3.35	0.26	0.1171	0.0871	0.067	0.0072	0.0005
T5	Alpha naphthyl acetic acid (NAA) 20 ppm	0.14	0.81	2.96	4.77	3.81	2.87	2.24	12.97	25.61	9.13	1.98	0.119	0.1173	0.0646	0.0133	0.0026
T6	Mepiuat chloride 5 % AS (5 %)	0.14	0.74	2.36	4.43	3.22	2.28	2.19	11.31	23.92	4.07	1.84	0.1146	0.1104	0.0665	0.0068	0.0029
T7	Brassinosteroid (25 ppm)	0.15	0.76	2.63	4.71	3.73	2.79	2.02	13.04	24.76	4.45	1.97	0.1162	0.123	0.0637	0.007	0.0029
T8	Water	0.14	0.71	2.06	4.12	2.84	2.01	2.11	8.39	15.73	2.65	0.23	0.1236	0.098	0.0596	0.0063	0.0005
T9	Control	0.13	0.59	1.42	3.32	2.48	1.89	1.87	7.83	15.47	2.58	0.13	0.1188	0.1	0.062	0.0064	0.0003
	Mean	0.14	0.72	2.25	4.29	3.17	2.29	2.17	10.56	20.98	4.36	1.11	0.1171	0.107	0.0634	0.0078	0.0017
	Sed	0.08	0.03	0.17	0.1	0.16	0.29	0.49	0.32	0.59	0.59	0.02	0.0016	0.0064	0.0098	0.0007	0.0005
	C.D (0.05)	NS	0.06	0.37	0.21	0.35	NS	NS	0.68	1.26	1.27	0.04	0.0036	0.0136	NS	0.0015	0.001

7.3.3 Biochemical Parameters

Apart from morphological and physiological characters, growth regulators also known to influence different biochemical parameters. The influence of different growth promoting and growth retarding substances, in Rabi soybean, in comparison to control and water spray on various biochemical parameters, namely, chlorophyll content, fluorescence emission and photosynthetic rate is discussed hereunder.

The application of NAA (20 ppm), brassinosteroid (25 ppm), mepiquat chloride 5 % AS (5 %) and chlormequat chloride 50 % SL at different concentrations had all resulted in significantly higher chlorophyll content, compared to control and water spray (Table 7.4). The high chlorophyll content noticed with the application of NAA was attributed to the protection of chlorophyll molecule from photo oxidation and increased chlorophyll synthesis. Further, [16] explained that application of mepiquat chloride to groundnut crop resulted in high chlorophyll content due to delayed chlorophyll degradation. Dong et al. [17] in soybean also reported the positive influence of brassinosteroid application on chlorophyll content, similar to findings of the present investigation. Similarly, increased chlorophyll content with the application of chlormequat chloride was reported earlier by Shalaby [18]. The fluorescence emission values (Table 7.4), namely, Fm, Fo, Fv and Fv/Vm and photosynthetic rate at flowering did not record any significant difference for the different growth regulator treatments studied in the present investigation.

In the present study, a significant increase in the seed protein content (Table 7.4) was noticed with the application of NAA (20 ppm), brassinosteroid (25 ppm), mepiquat chloride 5 % AS (5 %) and chlormequat chloride at different concentrations, compared to control and water spray indicating that, the applied growth regulators had marked effect on Biosynthetic pathways related to protein synthesis. Similar enhancement in seed protein content with the application of NAA [9], brassinosteroid [16], mepiquat chloride [9, 16] and chlormequat chloride [9] were also reported earlier.

The data pertaining to Nitrogen harvest index is presented in (Table 7.4). It was noticed to range from 16.26 (Control) to 17.78 (chlormequat chloride 50 % SL applied at 187.5 a.i/ha). A perusal of these results however, revealed the existence of non-significant differences between the treatments.

Table 4 Effect of different growth regulators on chlorophyll (SPAD-502), fluorescence, photosynthetic rate values, seed protein (%) and nitrogen harvest index in Rabi soybean

	Treatments	SPAD-502 value	Fo	Fm	Fv	Fv/Fm	Photosynthetic rate (μ mol CO_2 m^{-2})	Seed protein (%)	Nitrogen harvest index
T1	Chlormequat chloride 50 % SL (137.5 g a.i/ha)	32.74	59.67	173	115.91	0.67	17.5	41.8	17.06
T2	Chlormequat chloride 50 % SL (162.5 g a.i/ha)	32.6	61	179.3	116.54	0.65	16.44	42.17	17.31
T3	Chlormequat chloride 50 % SL (187.5 g a.i/ha)	32.53	62	184.3	123.48	0.67	16.98	42.63	17.78
T4	Chlormequat chloride 50 % SL (375 g a.i/ha)	31.3	62.67	182.3	122.14	0.67	16.36	41.2	16.49
T5	Alpha naphthyl acetic acid (NAA) 20 ppm	33.74	65.33	188	125.96	0.67	15.52	42.47	17.66
T6	Mepiuat chloride 5 % AS (5 %)	32.6	62.67	184.3	121.63	0.66	16	41.7	17.05
T7	Brassinosteroid (25 ppm)	33.03	65.67	188	124.08	0.66	16.36	42	17.29
T8	Water	30.26	63.67	179.67	120.37	0.67	15.82	40.43	16.44
T9	Control	30.16	61.67	177.67	117.26	0.66	15.88	40.2	16.26
	Mean	32.11	62.71	181.84	120.82	0.66	16.32	19.62	16.73
	Sed	0.3	1.78	6.91	4.52	0.29	1.64	0.14	1.26
	C.D (0.05)	0.63	NS	NS	NS	NS	NS	0.31	NS

7.4 Conclusion

The present study on Effect of Plant Growth regulators on Morphological, Physiological and Biochemical parameters of Soybean (*Glycine max* l. Merrill) revealed the superiority of NAA (20 ppm) and brassinosteroid (25 ppm) for majority of the morphological, physiological, biochemical, parameters for rabi soybean, compared to control, water spray and the growth retardants studied.

References

1. Association of official analytical chemists: AOAC (1980) Official methods of analysis, 12th edn. William Star Wet Glad (edn), Washington
2. Panse V, Sukhatme PV (1989) Statistical method for agricultural workers. ICAR, New Delhi, p 108
3. Lakshmamma P, Rao IVS (1996) Influence of shading and naphthalene acetic acid (NAA) on yield and yield components in black gram (*Vigna mungo* L.). Ann Agric Res 17:320–321
4. Shukla KC, Singh OP, Samaiya RK (1997) Effect of foliar spray of plant growth regulator and nutrient complex on productivity of soybean var. JS 79–81. Crop Res 13:213–215
5. Govindan K, Thirumurugan V, Aruchelvan S (2000) Response of soybean to growth regulators. Res Crops 1(3):323–325
6. Deotale RD, Katekhaye DS, Sorte NV, Raut JS, Golliwar VJ (1995) Effect of TIBA and B-Nine on morpho-physical characters of soybean. J Soils Crops 5:172–176
7. Rajesh K (2010) Influence of growth promoting and retarding compounds on growth, dry matter production and yield in green gram during *rabi*
8. Das A, Prasad R (2004) Effect of plant growth regulators on green gram (*Phaseolus radiatus*). Indian J Agri Sci 74(5):271–272
9. Girisha 2010 A comparative study on the growth promoting and retarding compounds on dry matter production and yield in black gram during rabi. M.Sc.(Ag) thesis submitted
10. Merlo D, Soldati A, Keller ER (1987) Influence of growth regulators on abscission off lower and young pods of soybeans. Eurosaya 5:31–38
11. Chandrasekar CN, Bangarusamy U (2003) Maximizing the yield of mung bean by foliar application of growth regulating chemicals and nutrients. Madras Agric J 90(1–3):142–145
12. Kumar P, Hiremath SM, Chetti MB (2006) Influence of growth regulators on dry matter production and distribution and shelling percentage in determinate and semi determinate soybean genotypes. Legume Res 29(3):191–195
13. Kothule VG, Bhalerao RK, Rathod TH (2003) Effect of growth regulators on yield attributes, yield and correlation coefficients in soybean. Ann Plant Physiol 17(2):140–142
14. Kumbhare MD, Khawale VS, Rajput GR, Datey CP, Dapugeanti KG (2007) Effect of nitrogen levels and chlormequat on mustard (*Brassica juncea* L.). J Soils Crops 17(2):394–397
15. Patil SB (1994) Effect of population levels and growth retardants on growth, yield and yield attributes and quality of soybean. M.Sc. (Agric) thesis, University of Agricultural Sciences, Dharwad
16. Jeyakumar P, Thangaraj M (1998) Physiological and biochemical effects of mepiquat chloride in groundnut (*Archis hypogaea*). Madras Agric J 85:23–26
17. Dong DF, Li YR, Jiang LG (2008) Effects of brassinosteroids on photosynthetic characteristics in soybean under aluminum stress. Acta Agron Sin 34(9):1673–1678
18. Shalaby MAF (2000) Influence of cycocel (2-chloroethyl ammonium chloride) on the vegetative growth, photo synthetic pigments, flowering, abscission and yield of faba bean (*Vicia faba*, L.). Ann Agric Sci Moshtohor, 38(3):1485–1502

Chapter 8
Rapid Diagnostic Tests Show False Positive Leading to Dilemma in Malarial Treatment: A Case Study

Susanta Kumar Panda and Amit Kumar

Abstract The absolute necessity for rational therapy in the face of rampant drug resistance places increasing importance on the accuracy of malaria diagnosis. Giemsa microscopy and Triple test method Rapid Diagnostic Test (RDT) represent the two diagnoses most likely to have the largest impact on malaria control in this present study. These two methods have their characteristic strengths and limitation. These tests were carried out on 156 patients of endemic areas of Gajapati district of Odisha State. These tests were evaluated by two methods, i.e. microscopic slide test and RDT kit method. RDT kits belong to Advantage MAL CARD Malaria pLDH (Plasmodium Lactate Dehydrogenase) antibody Pre-coated (J. Mitra & Co, India) card test, ParaHIT PfHRP2 (Histidine-Rich Protein 2) antibody pre-coated dip stick (Span Diagnostic Ltd, India) and S.D BIOLINE pf/pv capture antigen MSP (Merozoite Surface Protein) pre-coated card (S.D. BIO Standard Diagnostic Pvt. Ltd, India). A little amount of whole blood (5 μl) was taken using the plastic loop given with the kit, at the same time two slides are also taken. Out of 156 patients it was detected that 128 were negative and 28 were positive according to microscopic detection. Out of these 28 positives 23 were only *P. falciparum*, 2 were only *P. vivax*, 3 were having both PF and PV infection detected. On the other hand 32 were positive and 4 were false positive as shown by Advantage MAL CARD and ParaHIT, where as 49 were positive and 21 were false positive by the S.D. MSP pre-coated antigen card test kits.

Keywords Malaria · Malarial treatment · Rapid diagnostics test

S. K. Panda (✉) · A. Kumar
BioAxis DNA Research Centre (P) Limited, Hyderabad 500 068, Andhra Pradesh, India
e-mail: sushant@dnares.in

A. Kumar, *Biotechnology and Bioforensics*, Forensic and Medical Bioinformatics,
DOI: 10.1007/978-981-287-050-6_8, © The Author(s) 2015

8.1 Introduction

Malaria remains one of the most important endemic disease threats facing the all over the India. Lots of effort under taken to fight the disease and reduce the incident. But surprisingly day to day it becomes the king of the disease. So many methods employed in current malaria control programme with man, mosquito and malaria. Currently high quality researches in world wide simultaneously, high thought foot vector control programmes, community awareness and advance super susceptibility drugs to control the malaria, Education of malaria morbidity and drug resistance intensity plus the associated economic loss of these two factors require urgent scaling up of the quality of parasite-based diagnostic methods. An investment in anti-malarial drug development or malaria vaccine development should be accompanied by a parallel commitment to improve diagnostic tools and their availability to people living in malarious areas and development in both government and privet corporate sector. Now it needs social response and effort to all the community and serving employer to stop the life killing disease. The standard method of investigation in most medical units worldwide is still microscopic examination of blood films it is one of the gold standard method. Now newer methods are being utilised such as immunochromatographic detection of circulating parasite antigen and antibody (RDT) [1], Malaria Antigen enzyme-linked immunosorbent assay (ELISA) kit, detection of malaria antibodies by indirect immunofluorescence antibody assay (IFA), Methods for detecting malaria parasites by fluorescent staining also emerged (e.g., by the quantitative buffy coat (QBC) analysis, interference filter system for acridine orange-stained thin blood smear, flow cytometry, Polymerase Chain Reaction testing for plasmodium DNA, RT-PCR, LAMP (Loop-mediated isothermal DNA amplification). However, most of these techniques are complex, cost-effective and need specially trained technicians. Whereas rapid diagnostic test kits for malaria exist, which are fast, easy to perform, and can be carried out by relatively unskilled staff. The most commonly used tests for *P. falciparum* are based on the immuno-chromatographic detection of the Histidine-rich protein-2 (HRP-2), a protein produced by asexual stages and young gametocytes of *P. falciparum* [2], *Plasmodium* lactate dehydrogenase, PLDH [2, 3] can be either species-specific antibody detecting *P. falciparum* or *P. vivax* or 'pan-malarial' pLDH, detecting all four species of *Plasmodium* [4]. In addition, there is another antibody-coupled gold conjugate Aldolase which can detect all species of *Plasmodium* [5, 6]. Merozoite surface protein (MSP) is a recombinant pf/pv capture Antigen. Light microscopy remains the gold standard for malaria diagnosis [7], because it can provide information on both the species and parasite density of infection. But unfortunately, the procedure is not often used at the periphery of the health-care system in endemic because of the need for very skilled personnel for Malaria Parasite detection, power supply, good quality microscopes, handsome smear fixing and staining. Microscopy is also time-consuming and burdensome in remote area during epidemiological surveys where repeated measurements are required. Thus, there is an urgent need to develop a more sensitive and less expensive diagnostic test capable of rapidly detecting parasite antigen in blood at point of care. Availability of such diagnostic tools together with the use of artemisinin based combination therapy would be valuable in protection of special high-risk groups of children and pregnant women and delaying of development of resistance. In this study, we used four methods to detection of malarial parasite, (1) *Plasmodium Lactate dehydrogenase* (pLDH) Antigen detection in Blood sample of malaria patients using the Advantage MAL CARD malaria antibody Card (RDT) (J. MITRA & CO Pvt. Ltd, India) [8]. (2) PARAHIT *PfHRP*2 Dip Stick Span Diagnostic, (3) S.D BIOLINE pf/pv capture antigen *MSP* pre-coated card detecting antibody (S.D. BIO Standard Diagnostic Pvt. Ltd, India) card test. (4) Light microscopy with Gimsa staining this method has the advantage of high sensitivity, quantifiable results, and accurate speciation, but is fairly time-consuming and requires well-trained microscopists in order to detect low parasitemias and to properly differentiate the species [9]. Commercially available rapid diagnostic immuno capture test strips now exist which do not require the same level of training and equipment as microscopic examination, and are also significantly faster. However, as this review delineates, clinical trials have shown that the strips Or Card tests are not as sensitive as microscopic examination in detecting low level parasitemias, cannot quantify the level of malaria infection, and, at present, can only differentiate between *falciparum* and non-*falciparum* malaria [10, 11]. The strips also have problems relating to antigen persistence in the blood after parasite clearance from chemotherapy, leading to false positive post-therapeutic diagnoses, the test strips are currently not recommended to be used without a parallel blood smear sample being examined [12, 13]. By the overall study periphery serving (working) physician doctors are in dilemma how to treat the patient using RDT Kit.

8.2 Materials and Methods

8.2.1 Materials

RDTs: Advantage MAL CARD Malaria pLDH antibody Pre-coated (J. Mitra & Co, India) card test, ParaHIT PfHRP2 antibody pre-coated dip stick (Span Diagnostic Ltd, India) and S.D BIOLINE pf/pv capture antigen MSP pre-coated card (S.D. BIO Standard Diagnostic Pvt. Ltd, India). Light Microscopy using Gimsa Staining with Thick and Thin smear.

8.2.1.1 Method

This study was conducted between November 2012 and June 2013 at the Malaria endemic Community Health Centre of Rayagada and Gumma Block, Gajapati District of ODISHA STATE. As part of an ongoing anti malarial molecular MDR1 study and Drug Designing. Ethical clearance was obtained from the local Ethics

Committee of the District Malaria Office. Patients of all ages with suspected malaria were recruited according to routine criteria of the Malaria treatment (i.e., fever or a history of fever and/or other complaints indicating a possible malaria infection). Persons who came for follow-up visits of an earlier episode of malaria or within 4 weeks after a (confirmed and treated) malaria infection were excluded. Patients were asked for their informed consent and when accepted, they had their blood sampled for blood slides and 3 RDTs. Patients whose results were positive for malaria (for any test) were treated according to the National Protocol. A number of 156 patients had to be tested. The rapid test kits were opened only after the patient had been selected and interviewed by the technical person. Capillary blood was collected by finger-prick, sampling a standard volume of blood for each test according to the manufacturer's instructions, with the sampling device provided. Each selected patient had his/her blood examined by four methods requiring one drop of blood (5 μl) to be collected with a loop-shaped plastic sampling tool included with the device; there is one test line and other control line that demonstrates *P. falciparum* infection when these both turn pink and results are read at 15–20 min. Antibody individually packed dipstick kits detecting HRP2, Antibody detecting parasite pLDH specific for all species and MSP Antigen capture one region recombinant pf with recombinant pv capture antigen. Blood sampling of 5–10 μl is applied with a plastic capillary pipette then Buffer reagent clears the strip in about 10 min until control and/or test lines appear as pink coloured bands in a reading window, provided by manufacturer's instructions, so results are read after 20 min. Card-type test with one capture line specific for *P. falciparum* through *Pf* HRP-2 detection and the second line detecting all *Plasmodium vivax* species. Rapid diagnostic tests were read by the same bacteriologist and confirmed by a second independent reader when needed, all according to the manufacturer's instructions. The first person performed, read and recorded the results of the three tests and after that a second opinion was obtained from a second person reading again the same tests and recording the results. Each person read the RDT without knowing the result of the other reader or of the blood film. Results were compared and discussed to come to a consensus in case of different readings. At the end of this procedure, results were recorded on the patient's individual record form.

Microscopy Examination. Two thick smears were taken on one slide and one thin smear on a separate slide. Thick smears were submerged in methylene blue for 1 s, washed with Buffer solution and left to dry, thereafter stained horizontally with Field solution (one drop of solution A and one drop of solution B per 10 μl) in phosphate buffer B for 10 min, in accordance with nationally standard methods. Thin smears were fixed with methanol but not stained until necessary for species determination or better examination of the slide. Thick smears were evaluated by a well-trained, experienced microscopist, unaware of RDT results. A thick smear was considered negative if no parasites were seen in at least 200 fields. For positive smears, the number of parasites was counted in the number of fields needed to reach 200 white blood cells (WBCs) or 500 WBCs for low densities. Parasite density per micro litre was calculated assuming a standard of 8,000 WBCs per microlitre of blood as per WHO guidelines. Thin smears were used for species verification.

8.3 Results

The total blood samples investigated were 156. The results obtained revealed that, 28 were positive and 128 were negative by microscopy, 49 were positive and 107 were negative, out of 49 positives 21 were false positive by the SD Standard Diagnostic MSP Antigen test method. On the other hand 32 were positive and 124 were negative, 4 were false positive by (ParaHIT and Advantage MALCARD, HRP2 and pLDH). The results yielded by (SD Ag MSP) test demonstrated 86.5 % sensitivity, 90.2 % specificity and 78.1 % reliability. While Advantage MAL CARD (J. Mitra & Co, India) and PARAHIT (Span Diagnostic) HRP2 results revealed 98.2 % sensitivity, 98.9 % specificity and 97 % reliability.

8.4 Discussion

In this study, the SD Standard Diagnostic Ag MSP test sensitivity, specificity and reliability found to be less than Advantage MALCARD pLDH and PARAHIT Span Diagnostic HRP2 test for monitoring antimalarial's efficacy and the sensitivity of PLDH test during the follow up days showed that the test could have an impact in the diagnosis of *P. falciparum* malaria because of the high sensitivity obtained in the (98 %), with a considerable density of parasitaemia, of >100 parasites per micro litre of blood and in the negative cases as demonstrated by this study. PLDH test results obtained by this study indicated that the pLDH had a high sensitivity, specificity and reliability in Diagnosis antimalarials, correlates with the findings of the microscopy. However, it missed 4 cases that were positive by microscopy. HRP2 and PLDH test has the ability to monitor the success of chemotherapy and identifying drug resistant infections. False positive RDT results occur in a few percent of tests, Cross-reactivity with rheumatoid factor in blood generates a false positive test line, but replacement of IgG with IgM in recent products reduces this problem [14–16]. Cross-reactivity with heterophile antibodies may also occur.20 Occasional false negative results may be caused by deletion or mutation of the HRP2 gene [17]. It has been suggested that anti-HRP-2 antibodies in humans may explain why some tests were negative despite significant parasitemia [18]. Presence of an inhibitor in the patient's blood preventing development of the control line is also noted [19].

Quality control. For internal quality control (QC) and quality assurance (QA), a second independent reading was done by a different microscopist on about one third of the slides, especially low-density parasitemias and mixed infections. In cases in which both reference laboratories agreed on one diagnosis different from ours, results were corrected accordingly. Several factors in the manufacturing process as well as environmental conditions may affect RDT performance [20]. Manufacturers usually recommend 4–40 °C as the optimal temperature range. In practice, exposure of RDTs to >70 % humidity and/or >30 °C frequently occurs in the tropics. QC/QA measures are important to ensure that the purchased products meet performance expectations and that product quality is maintained through the delivery process to the periphery of the healthcare system. The recently introduced WHO initiative of RDT product testing and QA aims to standardize testing of RDTs and to assist countries and manufacturers with distribution and use [21]. Recommended guidelines for the field evaluation of malaria RDTs are available by WHO web Server.

8.5 Conclusion

In conclusion, at present, rapid diagnostic assays for malaria do not appear to be superior to blood smear examination the new generations of non microscopic immune chromatographic assay offer a practical chance to move the diagnosis of malaria away from the laboratory and nearer to the patient. However, MSP Ag was not showing high specificity and sensitivity versus pLDH, HRP2 antibody, beyond the clearance of peripheral parasitemia in certain cases reduces the usefulness of the ICT Malaria Pf/Pv test for monitoring the response to therapy and thus it was suggested that the disappearance of the parasite-specific enzyme pLDH and HRP2 after treatment may make the pLDH assay useful in predicting treatment. Although improvements in quantification of current antigens may improve predictive ability of treatment failure, the use of alternative antigens with more rapid clearance and greater sensitivity and specificity for viable parasites is essential for reaching a higher level of clinical utility. While currently available test cards or strips do have the advantage of a more rapid turn around and portability, the assay sensitivity, antigen persistence, and storage temperature limitations remain hindrances to these assays replacing the blood smear. As these features improve, there will be an increased advantage to using rapid test strips for malaria diagnosis.

Acknowledgments We acknowledge the Chief District Medical Officer, Department of Health, District Gajapati and kind cooperation of Community Health Centres medical officers, health workers and officials, for their help in Blood sample collection, Patient data collection, processing and analysis.

References

1. Am J Trop Med Hyg (2009) Performance of malaria rapid diagnostic tests as part of routine malaria case management in Kenya. Am J Trop Med Hyg 80:470–474
2. Piper R, Lebras J, Wentworth L, Hunt-Cooke A, Houze S, Chiodini P, Makler M (1999) Immunocapture diagnostic assays for malaria using *Plasmodium* lactate dehydrogenase (pLDH). Am J Trop Med Hyg 60:109–118
3. Palmer CJ, Lindo JF, Winslow I, Klaskala I, Queseda JA, Kaminsky R, Baum MK, Ager AL (1998) Evaluation of the optimal test for rapid diagnosis of *Plasmodium vivax* and *Plasmodium falciparum* malaria. J Clin Microbiol 36:203–206

4. Tjitra E, Suprianto S, Dyer M, Currie BJ, Anstey NM (1999) Field evaluation of the ICT malaria P.f/P.v immunochromatographic test for detection of *Plasmodium falciparum* and *Plasmodiumvivax* in patients with a presumptive clinical diagnosis of malaria in eastern Indonesia. J Clin Microbiol 37:2412–2417
5. World Health Organisation (2004) World health report 2004. World Health Organisation, Geneva, Switzerland
6. Moody A (2002) Rapid diagnostic tests for malaria parasites. Clin Microbiol Rev 15:66–78
7. Standard Diagnostics. One step malaria P.f/P.v antigen rapid test: SD bioline malaria antigen. Available at: http://www.standardia.com/insert/rapid/malaria_ag.htm. Accessed 20 Nov
8. Iqbal J, Hira PR, Sher A, Al-Enezi AA (2001) Diagnosis of imported malaria by *Plasmodium* lactate dehydrogenase (pLDH) and histidine-rich protein 2 (PfHRP-2)-based immunocapture assays. Am J Trop Med Hyg 64:20–23, Microbiol Rev 15:66–78
9. Humar A, Harrington MA, Pillai D, Kain KC (1997) *Para*Sight-F test compared with the polymerase chain reaction and microscopy for the diagnosis of *Plasmodium falciparum* malaria in travelers. Am J Trop Med Hyg 56:44–48
10. Am J Trop Med Hyg (2012) False-negative rapid diagnostic tests for malaria and deletion of the histidine-rich repeat region of the hrp2 gene. Am J Trop Med Hyg 86:194–198
11. WHO (2003) Malaria rapid diagnostics: making it work. Meeting report 20–23 Jan 2003, World Health Organization (RS/2003/GE/05[PHL]), Manila
12. Shiff CJ, Premji Z, Minjas JN (1993) The rapid manual para-sight-f test. A new diagnostic tool for *Plasmodium falciparum* infection. Trans R Soc Trop Med Hyg 87:646–648
13. Coleman RE, Maneechai N, Rachaphaew N, Kumpitak C, Miller RS, Soyseng V, Thimasaran K, Sattabongkot J (2002) Comparison of field and expert laboratory microscopy for active surveillance for asymptomatic *Plasmodium falciparum* and *Plasmodium vivax* in Western Thailand. Am J Trop Med Hyg 67:141–144
14. Laferl H, Kandel K, Pichler H (1997) False positive dipstick test for malaria. N Engl J Med 337:1635–1636
15. Grobusch MP, Alpermann U, Schwenke S, Jelinek T, Warhurst DC (1999) False-positive rapid tests for malaria in patients with rheumatoid factor. Lancet 353:297
16. Mishra B, Samantaray JC, Kumar A, Mirdha BR (1999) Study of false positivity of two rapid antigen detection tests for diagnosis of *Plasmodium falciparum* malaria. J Clin Microbiol 37:1233
17. Biswas S, Tomar D, Rao DN (2005) Investigation of the kinetics of histidine-rich protein 2 and of the antibody responses to this antigen, in a group of malaria patients from India. Ann Trop Med Parasitol 99:553–562
18. Durand F, Faure O, Brion JP, Pelloux H (2005) Invalid result of Plasmodium *falciparum* malaria detection with the Binax-NOW Malaria rapid diagnostic test. J Med Microbiol 54:1115
19. Peeling RW, Smith PG, Bossuyt PM (2006) A guide for diagnostic evaluations. Nat Rev Microbiol 4(Suppl):S2–S6
20. World Health Organization (2005) Malaria light microscopy. Creating a culture of quality. Report of WHO SEARO/WPRO workshop on quality assurance for malaria microscopy, Kuala Lumpur, Malaysia: RS/2005/GE/03(MAA), 18–21 April 2005
21. Bell D, Wongsrichanalai C, Barnwell JW (2006) Ensuring quality and access for malaria diagnosis: how can it be achieved? Nat Rev Microbiol 4:682–695

Chapter 9
Association of BDNF Levels and Muscoskeletal Problems in Type 2 Diabetes

Allam Appa Rao, Amit Kumar, Surendra Babu, Anuradha Parihar and Subha Senkhula

Abstract Brain Derived neurotrophic factor (BDNF) is very well reported in development of neurons and it plays major role in memory and interpretation. It is evident that BDNF is involved in maintaining the equilibrium of body weight and glucose homeostasis mechanism. Out of 96 Subjects included in this study Plasma BDNF level was found low in the patients with Type 2 Diabetes. We didn't find positive indications wrt association of BDNF G196A (Val66Met) polymorphism in diabetes or obesity. Type 2 Diabetic patients with the complaint of Joint pains were found to have even lower Plasma BDNF levels as compared to the diabetic patients without any Neurological problems or joint pains. Worsening BDNF Levels may be an alarming factor for type 2 diabetic patients with respect to development of Neurological disorders.

Keywords BDNF levels in diabetic patients · BDNF · Diabetes type 2

9.1 Introduction

A diet high in refined sugar minimizes the production of a crucial learning-related neurotransmitter called BDNF (Brain derived Neurotrophic factor) [1]. Our learning capacity diminishes in the absence of BDNF. As the level of BDNF

A. A. Rao (✉)
Jawaharlal Nehru Technological University, Kakinada, Andhra Pradesh, India
e-mail: apparaoallam@gmail.com

A. Kumar · S. Babu · A. Parihar
BioAxis DNA Research Centre (P) Limited, Hyderabad 500 068, Andhra Pradesh, India

S. Babu
Andhra University, Visakhapatnam, India

S. Senkhula
Jawaharlal Nehru Technological University, Vijayanagaram, Andhra Pradesh, India

A. Kumar, *Biotechnology and Bioforensics*, Forensic and Medical Bioinformatics,
DOI: 10.1007/978-981-287-050-6_9,

worsens, insulin resistance increases and patients develop Type 2 Diabetes. Neurotrophins help in controlling neurogenesis [2]. Majority of neurons in the brain of mammals are formed prenatally, Neurotrophins are the growth factors which have been reported to have the potential role in the survival [3], structural development and depicting the function of Neurons. One such neurotrophin BDNF which is expressed abundantly in brain has major effect on neurons via the tyrosine kinase receptors. Recent studies have shown the expression levels of BDNF in the skeletal muscles also [2]. Brain-derived neurotrophic factor (BDNF) and its tyrosine kinase receptor, TrkB, are also reported to have their association in body mass and weight regulation and their expression in hypothalamic nuclei which are directly connected to eating habits, amount of food intake etc. [4].

It is reported that men with low BDNF levels have less efficiency to perform complicated tasks as compared to the individuals with normal BDNF levels in the body. BDNF governs synaptic plasticity and is able to change neuronal excitability and synaptic transmission [5, 6]. Present study tested the hypothesis that patients with type 2 diabetes have low levels of BDNF in plasma.

9.2 Sampling

Participants were divided in the following categories

1. Healthy Non Smokers (22)
2. Healthy smokers (4)
3. Diabetic obese (18)
4. Diabetic smokers (2)
5. Diabetic non obese and non smokers (50)

We collected the samples of 96 subjects from in and around Hyderabad, India which included 70 males and 27 females. The patients involved in the study were reached by visiting the hospitals, multispeciality diagnostic centers, by phone calls, emails etc. Detailed medical information on type 2 diabetes of the subjects was gathered on the basis of oral information and medical test from each subject. During the discussion on the kind of the life style maintained by these patients, we even discussed regarding the Diabetic family history and the kind of complications because of this. We received mixed response on genetic history and nothing very substantial could be connected to the current diabetic conditions of the patients participated in this study.

One of the patients was Diabetic as well as suffered from cardio vascular disease; angioplasty was done for this case in 1996.

We excluded the patients reported with severe infections, depressions and dementia. All the subjects participated in the study were aged between 45 and 65 years, Asian, Indian. Patients provided the blood samples for genotyping as well as BDNF Quantification.

To verify the diagnosis, the WHO diagnostic criteria for type 2 diabetes were used. Participants were carefully screened to isolate metabolic conditions other than type 2 diabetes that are known to influence body composition and the immune system. Exclusion criteria were: treatment with insulin, recent or ongoing infection, history of malignant disease and known dementia. Participants were categorized as having cardiovascular disease (CVD) if they had claudication or at least one of the following diagnoses: cerebrovascular accident, angina pectoris, or prior coronary artery bypass graft or percutaneous transluminal coronary angioplasty. Current intake of antidepressive medication was registered. Participants received the complete oral information about the experimental procedure and aim of the study and signed the written consent form.

WHO diagnostic criteria was followed

Fasting Blood Glucose

3.9–5.5 mmol/l—Normal Glucose Tolerance
5.5–6.9 mmol/l—Impaired Glucose Tolerance
7.0 and above—Diabetes

OGTT

>7.8 mmol/l—Normal Glucose Tolerance
7.8–11.0 mmol/l—Impaired Glucose Tolerance
11.1 and above—Diabetes

9.3 Methodology[1]

After the general Physical examination of the subjects the Body mass index was calculated and Oral glucose tolerance test (OGTT) was performed on the samples drawn before and after taking Glucose. Blood Glucose and insulin were measured using routine laboratory methods. Blood samples were centrifuged at 4,000 RPM for 15 min at 4 °C, plasma was collected and stored at −20 °C. Plasma BDNF concentrations were measured by Promega BDNF Emax Immunoassay System. DNA was extracted from whole blood using BDRC DNA Extraction Kit, India and was stored at −20 °C. Quantification of isolated DNA was done by Spectrophotometry, Miostech UV 120, Australia. Amplification of BDNF gene was done on ABI 9,700 PCR System in a total reaction volume of 50 μl. The reaction mixture consisted of 2 μl of 50 ng/μl purified gDNA and 2 μl each of forward and reverse primers of 20 pmol/μl. The in silico primer designing for the reaction was done on

[1] Present study was approved by the Institutional Ethical Committee for Research on Human volunteers, Andhra University Visakhapatnam.

Graph 9.1 BDNF levels (Y axis) as compared to the Glucose levels (X axis)

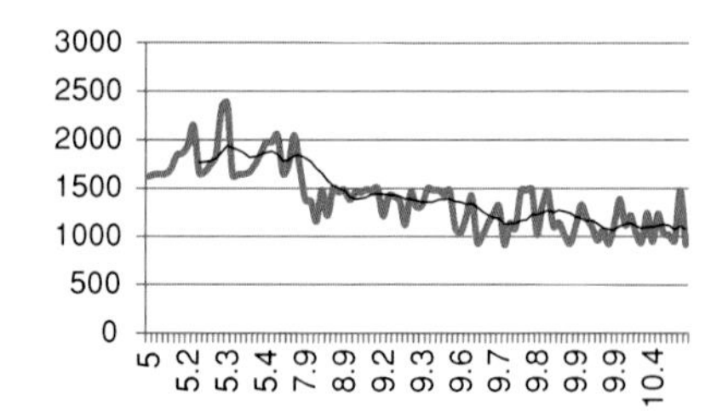

Table 9.1 G196A mutation

Condition	G196A	BDNF level (pg/ml)
Diabetic	Yes	1,235.45
Diabetic	Yes	1,209.42
NGT	Yes	1,643.28
NGT	Yes	1,858.33
NGT	Yes	1,739.2

Primer 3 [7]. [Cycle profile: 95 °C for 10 min + (95 °C for 1 min + 55 °C for 1 min + 72 °C for 1.30 min) × 35 cycles.] The samples were then subjected to SNP Validation for 196 G/A polymorphism using fluorescence-based Applied Biosystems Stepone™ Real Time PCR System. The probe used was C_11592758_10 containing forward and reverse primers of Applied Biosystem.

9.4 Result and Discussions

Prior studies on Animal models suggest that BDNF plays important role in insulin resistance [8]. BDNF reduces food intake and in turn helps in lowering blood glucose levels in obese diabetic mice [9]. Another study suggests that synaptic responses are also governed by the Neurotrophins [10]. It is a proven fact that age has impact on the BDNF Levels too [11]. In this study Plasma BDNF Levels were found low in the Type 2 diabetic patients. As per the results obtained it may be concluded that Glucose levels in our body is inversely proportional to the BDNF Levels in Diabetic patients (Graph 9.1). The patients who reported Joint pains along with Type 2 diabetes had even lower BDNF Levels. G196A Polymorphism has been reported to have association with the Neurological problems like Alzhimers but some of the studies contradict it too [12–14].

We couldn't find any sound correlation between BDNF G196A (Val66Met) polymorphism and its association with increased sugar levels or type 2 diabetes or obesity onset. It was reported both in Healthy (NGT) as well as Diabetic Patients (Table 9.1).

The BDNF levels were found to be independent of sex in Type 2 Diabetes. Generally Plasma BDNF Levels were found independent of the smoking criterion.

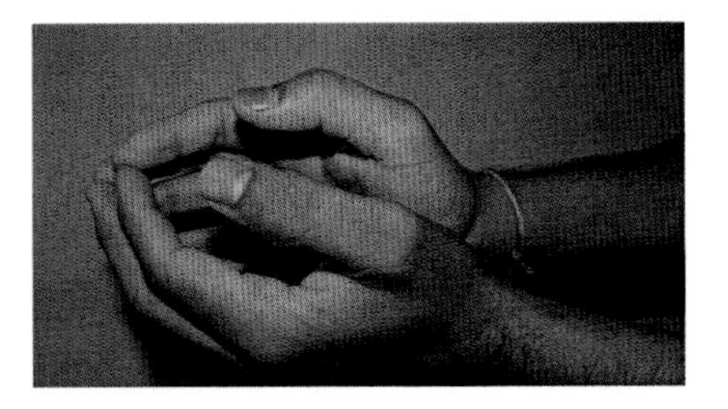

Fig. 9.1 Diabetic patient with joint pain in hand

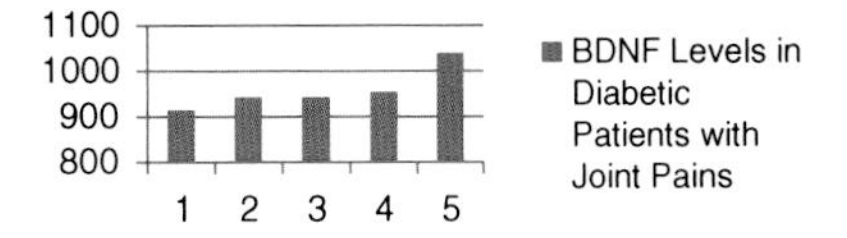

Graph 9.2 BDNF levels in diabetic patients with joint pains

A slight increase in the BDNF levels was reported in 2 smoking patients which may be a case of nicotine dependence pathophysiology [15].

Diabetes affects the musculoskeletal system in a variety of ways. The metabolic perturbations in diabetes (including glycosylation of proteins; microvascular abnormalities with damage to blood vessels and nerves; and collagen accumulation in skin and periarticular structures) result in changes in the connective tissue [16].

In our study 6 out of 96 patients complained of Joint pains which included 4 males and 2 females. 3 of them had the problem in the pedal bones 1 in the ankle and other 2 in the wrist and fingers of hand. The patients suffered with the joint problems were suffering from Type 2 diabetes for more than 10 years. One male Patient with Joint pain in hand showed inability to join both the palm completely while pressing them against each other, there was a clear and visible gap between the two palms (Fig. 9.1). While discussion with the patients most of them said that the problem of joint pains developed very slowly which may have progressed because of age along with the Diabetes.

Diabetic Patients Complaining the Joint pains had a maximum BDNF level of 1,085 pg/ml (1 Patient) which is even lower Plasma BDNF level as compared to the diabetic patients who didn't develop Joint pain (Graph 9.2).

9.5 Conclusion

There were three main conclusions derived from this study, (1) BDNF levels are low in the patients with Type 2 Diabetes. (2) There is no correlation of G196A polymorphism in Type 2 Diabetes. and (3) An association with the BDNF Levels and the muscoskeletal problems. Further studies on the connection of BDNF Levels with Neurodegenerative problems may open doors to maintain the patients developing neurological problems after Diabetes type 2.

Funding and Support This Research was solely funded by Department of Science and Technology, Government of India under IRHPA Scheme vide Lr No. IR/SO/LU/03/2008/1.

Conflict of Interest

Authors didn't have any conflict of interest in this Research.

References

1. Molteni R, Barnard RJ, Ying Z et al (2002) A high-fat, refined sugar diet reduces hippocampal brain-derived neurotrophic factor, neuronal plasticity, and learning. Neuroscience 112:803–814
2. Bath KG, Lee FS (2010) Neurotrophic factor control of adult SVZ neurogenesis. Dev Neurobiol 70(5):339–349 Special Issue: Trophic factors: 50 years of growth
3. Mattson MP, Maudsley S, Martin B (2004) BDNF and 5-HT: a dynamic duo in age-related neuronal plasticity and neurodegenerative disorders. Trends Neurosci 27:589–594
4. Matthews VB, Åström M-B, Chan MHS, Bruce CR, Prelovsek O, Åkerström T, Yfanti C, Broholm C, Mortensen OH, Penkowa M, Hojman P, Zankari A, Watt MJ, Pedersen BK, Febbraio MA (2009) Brain derived neutrophic factor is produced by skeletal muscle cells in response to contraction and enhances fat oxidation via activation of AMPK. Diabetologia 52:1409–1418
5. Kernie Steven G, Liebl Daniel J, Parada Luis F (2000) BDNF regulates eating behavior and locomotor activity in mice. EMBO J 19:1290–1300
6. Causing CG, Gloster A, Aloyz R, Bamji SX, Chang E, Fawcett J, Kutchel G, Miler FD (1997) Synaptic innervation density is regulated by neuronderived, BDNF. Neuron 18:257–267
7. Rozen Steve, Skaletsky Helen J (2000) Primer3 on the www for general users and for biologist programmers. In: Krawetz S, Misener S (eds) Bioinformatics methods and protocols: methods in molecular biology. Humana Press, Totowa, pp 365–386
8. Ono M, Ichihara J, Nonomura T et al (1997) Brain-derived neurotrophic factor reduces blood glucose level in obese diabetic mice but not in normal mice. Biochem Biophys Res Commun 238:633–637
9. Krabbe KS, Nielsen AR, Krogh-Madsen R et al (2007) Brain-derived neurotrophic factor (BDNF) and type 2 diabetes. Diabetologia 50(2):431–438
10. Figurov A, Pozzo-Miller LD, Olafsson P, Wang T, Lu B (1996) Regulation of synaptic responses to high-frequency stimulation and LTP by neurotropins in the hippocampus. Nature 381:706–709
11. Lommatzsch M, Zingler D, Schuhbaeck K et al (2005) The impact of age, weight and gender on BDNF levels in human platelets and plasma. Neurobiol Aging 26(1):115–123
12. Li Y, Rowland C, Tacey K et al (2005) The BDNF Val66Met polymorphism is not associated with late onset Alzheimer's disease in three case-control samples. Mol Psychiatry 10(809–810):26
13. Surtees PG, Wainwright NWJ, Willis-Owen SAG et al (2007) No association between the BDNF Val66Met polymorphism and mood status in a non-clinical community sample of 7389 older adults. J Psychiatr Res 41(5):404–409
14. Matsushita S, Arai H, Matsui T et al (2005) Brain-derived neurotrophic factor gene polymorphisms and Alzheimer's disease. J Neural Transm 112:703–711
15. Maggio R, Riva M, Vaglini F, Fornai F, Racagni G, Corsini GU (1997) Striatal increase of neurotrophic factors as a mechanism of nicotine protection in experimental Parkinsonism. J Neural Transm 104:1113–1123
16. Kim RP, Edelman SV, Kim DD (2001) Musculoskeletal complications of diabetes. Clin Diab 19(3):132–135